MATHEMATICAL THEORY OF STELLAR ECLIPSES

by

ZDENĚK KOPAL
Department of Astronomy,
University of Manchester, England

KLUWER ACADEMIC PUBLISHERS
DORDRECHT / BOSTON / LONDON

Library of Congress Cataloging in Publication Data

Kopal, Zdeněk, 1914-
    Mathematical theory of stellar eclipses / Zdeněk Kopal.
       p.   cm.
    Includes bibliographical references.
    ISBN-13:978-94-010-6729-4   e-ISBN-13:978-94-009-0539-9
    DOI: 10.1007/978-94-009-0539-9
    I. Title.
 QB421.K65   1990
 523.8'444--dc20   90-30622

ISBN-13:978-94-010-6729-4

Published by Kluwer Academic Publishers,
P.O. Box 17, 3300 AA Dordrecht, The Netherlands.

Kluwer Academic Publishers incorporates
the publishing programmes of
D. Reidel, Martinus Nijhoff, Dr W. Junk and MTP Press.

Sold and distributed in the U.S.A. and Canada
by Kluwer Academic Publishers,
101 Philip Drive, Norwell, MA 02061, U.S.A.

In all other countries, sold and distributed
by Kluwer Academic Publishers Group,
P.O. Box 322, 3300 AH Dordrecht, The Netherlands.

Printed on acid-free paper

MATHEMATICAL THEORY OF STELLAR ECLIPSES

LIST OF CONTENTS

Chapter I

INTRODUCTION

ASTRONOMICAL ECLIPSE PHENOMENA

In looking over the long history of human science from time immemorial to our own times, it is impossible to overestimate the role played in it by the phenomena of eclipses of the celestial bodies—both within our solar system as well as in the stellar universe at large. Not later than in the 4th century B.C., the observed features of the shadow cast on the Moon by the Earth during eclipses led Aristotle (384–322 B.C.) to formulate the first scientific proof worthy of that name of the spherical shape of the Earth; and only somewhat later, the eclipses of the Sun provided Aristarchos (in the early part of the 3rd century B.C.) or Hipparchos (2nd half of the same century) with the geometric means to ascertain the distance which separates the Earth from the Sun. In the 17th century A.D. (in 1676, to be exact) the timings of the eclipses of the satellites of Jupiter by their central planet enabled Olaf Römer to discover that the velocity with which light propagates through space is finite.

The total eclipses of the Sun have, in particular, proved a veritable godsend to students of many branches of science as well as humanities—from the historians of ancient times (whom they enabled to straighten out many an obscure date of early chronology), to the geophysicists (who used old eclipse records to detect secular irregularities in the length of the day); or again to the chemists who detected (in 1868) in the solar "flash spectrum" the emission lines of a new element—helium—before the latter was eventually detected in our terrestrial laboratories. Students of the apparent motion of the Moon in the sky found the historical records of ancient eclipses invaluable for an identification of its "secular acceleration" (more recently identified with the previously neglected perturbations of the Moon's motion caused by Jupiter and Saturn); and occultations, by the Moon, of the stars of accurately known positions continue to serve as tools to locate the exact positions of our satellite in the sky; while high-speed photometry of such occultations led, in several cases, to the determinations of apparent angular diameters of the respective stars more accurate than would (at present) be possible by any other alternative method. Such photometry of the occultations of background stars by Venus—or major planets—repeatedly served in recent years as valuable tools for an exploration of the structure of their extended atmospheres and, in at least two cases (Uranus and Neptune) led to the discovery of rings surrounding these planets before their existence was directly verified by spacecraft. Moreover, in recent decades the astrophysicists have also taken advantage of the fleeting minutes of total eclipses to measure the extent of the deflection of the images of background stars in the gravitational field of the Sun, and thus to study the metric properties of space in the neighbourhood of large masses.

But it is not only the eclipses involving the celestial bodies within our solar system that have proved to be so scientifically rewarding; for an even much greater prize was in store for us in the realm of the stars.

From time immemorial, the firmament of the heavens was regarded by the ancients as fixed and immutable (apart from a small number of "wandering stars" which turned out to be our sister-planets); and a recognition of changes within it antedate only shortly the discovery of the telescope. To be sure, occasional appearances of "guest stars" visible to the naked eye were noted centuries before; but were regarded as curiosities outside the well-established world order.

The first instance of a star which varied in brightness in the sky—that of omicron ("Mira") Ceti was, to be sure, detected in 1596 (by David Fabricius—once pupil and assistant of Tycho Brahe) shortly before the discovery of the telescope—eventually to become the prototype of "long-period variables" of which many hundreds have been discovered since that time (and not only in our Galaxy). It took, however, almost a century before an Italian astronomer—Geminiano Montanari of Bologna—happened to note (cf. Montanari, 1671) that another naked-eye star (β Persei, or Algol as it was called by the Arabs) does not always shine with constant light; and voluminous manuscripts he left behind enabled later scholars (e.g., Porro, 1891) to specify 8 November 1670 as the date of the first recorded minimum of Algol (though its hour, let alone minute, remain unknown).

Whether or not this was the first instance when the variability of Algol was noted still remains a moot point (for its discussion cf. Chapter I of the writer's treatise on *Close Binary Systems* (Kopal, 1959; or, more recently, Budding, 1988). However, it is certain that before the 17th century came to its close, the astronomers of that time knew of the existence of at least two—and probably three—variable stars in the sky (including χ Cygni—a long-period variable which (like Mira) becomes likewise visible to the naked eye near the time of its maximum, brightness; and detected as such by Gottfried Kirch in 1686). These first-known variable stars—all discovered at a time when our astronomical ancestors were still primarily concerned with positional astrometry—were destined to become prototypes of variables whose study, in fullness of time, contributed essentially to the outgrowth of stellar astrophysics in our own times. Of these, however, the principal concern of this book will be confined to variable stars of the Algol type; and it is to their history as it unrolled in the past three hundred years that we shall hereafter attempt to append some footnotes.

Although this history goes back to 1670—the year when the variability of Algol was discovered 60 years after the advent of telescopic astronomy—its first chapters did not unroll any too quickly; for almost another century had to elapse before Algol's light changes were recognized to be periodic.

This came to pass (as far as we know) in the autumn months of the year 1782—no longer under the clear Italian sky where Algol's variability was discovered, but in the cloudy climate of northern England, in the place the ancient Romans knew as Eboracum (and the modern tourists under the name of York); and the first minimum of Algol was observed there on 12 November 1782 (102 years after that observed at Bologna by Montanari). This should be a memorable date in the history of stellar astronomy—not only because from that time Algol would never be forgotten (Kopal, Plavec and Reilly, in their 1960 publication, listed the times of no less than 1537 minima of Algol observed between 1782 and 1950); but also because of the person of its discoverer. John Goodricke (1764–1786)—a scion of an old family of English gentry and an amateur astronomer *par excellence*—was only 18 years of age at the time of this discovery; but being severely handicapped (deaf-mute since childhood) and of infirm health which carried him away at 22, he was generously treated by history which need not be repeated in this place (for its fuller details cf., eg., Chapter 6 of a recent book by the present writer (Kopal, 1986) and other sources quoted therein).

A few additional comments may, however, be made which came to light in the very recent past. As was recently pointed out by Hoskin (1979), Goodricke's laurels for proposing that Algol's changes of light are caused by mutual eclipses of the components in a close binary system (cf. Goodricke, 1783) should also be shared with his close friend (and, in fact, a cousin) Edward Pigott (1753–1825) of the Fairfax family, who in the days of his youth shared Goodricke's astronomical interests. Unencumbered by any of the physical handicaps which made life difficult for his cousin, Pigott can also claim credit for independent discoveries of the variability of naked-eye star η Aquilae (in 1784; this eventually turned out to be a cepheid variable); and in the next decade he noted also (irregular) changes of light of R Coronae Borealis (1795), and of R. Scuti (1797) which proved to be semi-regular.

Goodricke was, however, luckier in some other respects: for in addition to the periodicity of Algol, [1] within his short life he found time to discover also the variability of two other naked-eye stars—β Lyrae (of period estimated by Goodricke to 13 days; its more exact value being 12.91d) which proved to be the second known eclipsing variable, as well as of δ Cephei itself—both in 1784. In fact, Goodricke's discovery of β Lyrae, and Pigott's of η Aquilae were made on the same night (20 September 1784); and δ Cephei was added to the list within one week! These must have been enchanted nights for both young astronomers; but, alas, their partnership was destined not to last. For as early as 1786 (20 April) John Goodricke passed away ("in consequence of a cold from exposure to night air during astronomical observations"; see Clegg, 1961); and the repercussions for Pigott to the early loss of his friend must be left to conjecture. But if this was what led him shortly thereafter to forsake astronomy altogether, wasn't this the last monument which Goodricke erected for himself in the history of our science?

---

[1] Goodricke's original value for Algol's period was 2 days, 20 hours and 45 minutes; differing from its true value by only 4 minutes. A year later Goodricke revised this period to 2 days, 20 hours, 49 minutes and 9 seconds—a result on which modern observations have improved but little!

But to return to Algol—new material brought to light in recent years by Hoskin at York leaves no room for doubt that John Goodricke and Edward Pigott—far from being only family relations—intellectually collaborated; and, therefore, the actual paternity of their ideas on the cause of Algol's variability may be difficult to disentangle after two hundred years. However—and this is essential—Goodricke did publish his views (with Pigott's knowledge) in 1783 for the benefit of his contemporaries and of posterity; while Pigott limited himself to recording them in his private diaries which remained unknown for almost 200 years, and but recently came to light (see Hoskin, 1979). According to the standards prevalent in the 18th century as well as today, Goodricke's priority of discovery should, therefore, be unquestioned—and it was not questioned (either by Pigott or his contemporaries) up to the present time. Moreover, soon after Goodricke died in April 1786, Pigott himself (although he had still almost 40 years to live) abandoned further pursuit of active scientific work—perhaps as a mute testimony of the fact that both needed each other to bring the best out of them—as they did in the 4 years of their fruitful collaboration; and their joint story ended, in effect, when Goodricke's body was laid to rest in a tomb at Hunsingore cemetery (close to the former family seat at Ribston Hall in Yorkshire), which was only recently re-discovered (cf. Kopal, 1986).

How did the views of those young astronomers—and, in particular, their atempt to account for Algol's periodic changes of light by eclipse hypothesis—fare in the opinions of their contemporaries? Truth be said, not very well; for they were clearly ahead of their time. Consider, for instance, the opinion of William Herschel, who apparently acted as a referee of Goodricke's paper submitted to the Royal Society of London in 1783: "The idea of a small Sun revolving around a large opake body has also been mentioned in the list of such hypotheses," wrote Herschel in a report read before the Royal Society on 8 May 1783 (but not printed till in *The Scientific Papers of Sir William Herschel*, London 1912; vol. I, p.cvii). He had reasons to doubt the correctness of Goodricke's hypothesis; for (as he wrote) prior to his report he had observed Algol repeatedly with his 7-foot telescope (of the discovery of Uranus fame); but each time found it to be "distinctly single". Twenty years later, it became Herschel's destiny to demonstrate the existence of physical binaries in the sky; but whether or not the ageing astronomer ever made up his mind about Algol we do not know.

Moreover, with the advent of the 19th century, the evolution of our subject accelerated but little: astronomers of that time continued to be preoccupied by measuring mainly the positions of the stars; and their instrumental equipment was lorded over by the meridian circle, sharing only grudgingly its prestige at the observatory with long-focus refractors. And what was true of the equipment, was equally true of those who manned it. Even towards the end of that century, Edward C. Pickering of Harvard College Observatory could be upbraided by his senior Harvard colleague, Benjamin O. Pierce, for wasting the observatory's time and resources on photometric surveys of the sky, when these could be better used to measure the positions of the stars with an accuracy exceeding that of photometric measures by at least four orders of magnitude.[2]

In such an atmosphere then prevalent in the professional world, it should perhaps come as no surprise that the number of known eclipsing variables grew but very slowly—in fact, not a single new one was discovered since Goodricke's death in 1786 until 1848, when Baxendall discovered the Algol-type variability of the naked-eye star λ Tauri, and Hind discovered the first totally-eclipsing system S Cancri (the eclipses in the Algol systems are partial). To be sure, prior to that time (in 1821) the light of another naked-eye star ϵ Aurigae was found to be variable, but in an enigmatic manner; and a hundred years had to elapse before this star too made history of its own.

To return to the 19th century, in the second half of it the situation began at last to change as a preview of greater things to come, and while only two eclipsing variables (Algol and β Lyrae) were known by 1800, by 1900 their number increased to 24. This latter total included an additional partially-eclipsing naked-eye star δ Librae (1858), followed by such well-known telescopic objects as the totally-eclipsing systems U Cephei (discovered by Ceraski in 1880) and U Sagittae (Schwab, 1901); or partially-eclipsing systems RS Sgr (by Gould, 1874), U Oph (Sawyer, 1881), R CMa (Sawyer, 1887) or Y Cyg (Dunér, 1892). In fact, more than one-half of 24 eclipsing stars known by the end of the 19th century were discovered during its last decade—due largely to the new (photographic) techniques introduced in practical use at that time; the first eclipsing variable so discovered (by Miss Wells of Harvard in 1895) was the totally-eclipsing system W Delphini.

However, it was not only the increasing stars detected by new techniques, but also parallel developments in astronomical spectroscopy, that gave a new dimension to the investigation of such celestial systems. Indeed, it was not till the spectroscopic work by H. C. Vogel in 1889 (cf. Vogel, 1890) on Algol—demonstrating that the moments of its minima of light coincided with the conjunction of its spectroscopic orbit—which confirmed with final validity Goodricke's double-star hypothesis; and the same has been found true of every eclipsing variable so tested since that time. Both these methods—photometric and spectroscopic—for testing the binary nature of the stars are, of course, not equally efficient from the observer's point of view: for with the aid of large reflectors which came into use since the beginning of this century, spectroscopic binaries can be detected (depending on their absolute magnitude) up to distances of several thousand parsecs; but beyond that domain (which means in a large part of our own Galaxy and, of course, in other members of our local metagalaxy) close binaries can be detected if—and only if— the orientation of their orbits in space is such that they can exhibit the characteristic variations of light caused by eclipses.

As a result of such work the number of known eclipsing binaries—2 by the

---

[2]Pickering (an MIT-trained physicist) was, however, unrepentant and justly so; for the international reputation of his observatory was soon based on Harvard Photometry and Henry Draper Catalogues; while (truth be said) the AG Catalogue zones for which he assumed responsibility soon turned out to be the least precise of the lot.

year 1800 and only 24 by 1900—has in the first half of this century exceeded two thousand (cf. Kukarkin and Parenago, 1948); though by the year 2000 it will no doubt exceed ten thousand. However, even so such numbers are only a measure of the zeal with which astronomers have been busy in this particular field of our vineyard. Within (approximately) 30 parsecs of the Sun we now know of at least 10 eclipsing variables among some 10000 individual stars; and if the same proportion extends also beyond this distance, the total number of eclipsing stars in our Galaxy alone should be of the order of 10^9—i.e., quite beyond the means of individual discovery. Moreover, such exotic variety has already been found among them that binary nature is now held responsible (possibly with justification) for most peculiarities exhibited by the stellar population as a whole; and it is likely that this tendency will only grow with time.

But the significance of eclipsing variables in astronomy is not based on their enormous numbers (disclosing them to be standard handiwork of nature), but also on the unique nature of information which they alone can provide. Spectroscopic observations alone can furnish only the minimum values of the masses of their components, and lower limits for the size of their orbits. The "missing clue"— necessary to convert these lower bounds into actual values—is the inclination of the orbital plane to the line of sight; and its value can be ascertained (from an analysis of the light changes) only if the respective binary happens also to be an eclipsing variable. Moreover, the range of astrophysical data which can be deduced from such an analysis transcends by far the masses and absolute dimensions of their components, or the characteristics of their orbits; for even the internal structure of the constituent components may transpire from some of their observable characteristics. Although the interiors of the stars are concealed from direct view by the enormous opacity of the overlying material, a gravitational field emanates from them which the overlying layers (opaque as they may be) cannot appreciably modify. This field is, to be sure, invisible; but is bound to affect both shape and motion of any masses that may be situated within it—just as the distribution of brightness over the exposed surfaces of the components (influencing the light changes within eclipses) is governed by the energy flux originating in the deep interior of the respective stars.

These, and many other possibilities opening up by the studies of eclipsing variables have long attracted due attention on the part of the observers. Largely because of the recurrent nature of the phenomena exhibited by such systems, eclipsing variables have always been favourites with pioneers of accurate photometry of any kind—visual, photographic, or photoelectric—and the total number of individual observations made in this field must by now run into many millions. A mere inspection of such data cannot, however, exhaust—or even scratch—the tremendous wealth of information which they contain. To develop this information requires the construction of a clue for deciphering the observed light changes, and with its aid to extract all significant information from the observations—no more and no less than the data contain. To summarize the methods by which this can be accomplished sets the principal task which we propose to outline in this book.

In order to limit ourselves to the essentials, we shall in what follows have but little to say on the observational basis of our problem; for this aspect has already been covered in several recent sources (cf., in particular, Kopal, 1979 and 1986). Indeed, the bulk of the contents of this volume naturally splits up in two distinct parts. The first (consisting of Chapters II–IV) will be concerned with the setting up of a photometric model of close binary systems whose components can eclipse each other and, in doing so, exhibit characterisric changes of light which it will be our task to analyze. Most of this subject matter is of very recent date—barely more than 15 years old—and its novelty largely rests on our regarding the observed loss of light as a cross-correlation of two distinct apertures of shapes which (for close binaries) vary with the phase; and the natural language for such a description of the eclipse phenomena will turn out to be the Hankel transforms. To some readers, such an approach may perhaps be less familiar than a more pedestrian (i.e., algebraic) approach followed by previous investigators; yet lesser familiarity should be more than offset by greater generality of approach to problems expressible in terms of functions which, by their nature, are bound to be discontinuous.

In the subsequent Chapter V we shall, however, embark on the discussion of a very different problem: while in Chapters II–IV our main aim has been to describe the photometric manifestations of close eclipsing systems as functions of the *time* (or phase) in terms of physically self-consistent models characterized by a certain group of the "elements of the eclipse", in Chapters V and VI we shall consider the *converse* problem of extracting these elements from the observations by a recourse to the "moments of the light curves" defining, in effect, a (continuous) *Fourier transform* of the observed light curves. As it transpired in the past 15 years (since Kopal, 1975), these "moments" correspond to frequency curves from which the original (observed) light curves can be synthesized; and the principal advantage of a translation of our problem from the "time-domain" is the fact that while the shape of the light curve is related with the elements of the system we aim to determine in a transcendental manner, its Fourier transform is related with them algebraically—a fact which renders our problem capable (among other things) of a rigorous error analysis. The elements of this new approach have already been expounded (and compared with the time-domain analysis) in a previous book by the present writer which appeared 10 years ago (cf. Kopal, 1979); the aim of the present volume is to bring the subject up to date on the basis of new work performed since that time.

From the progress achieved so far the reader can surmise for himself whether, and how fast, our road may lead us also in the future. In venturing on any such estimate, it is well to keep in mind what happened in the past. While the first known eclipsing variable—namely, Algol—was discovered by Geminiano Montanari at Bologna in November 1670, more than a century elapsed till John Goodricke detected (in November 1782) the periodicity of its light changes and, shortly thereafter (May 1783) published his famous hypothesis proposing binary nature as a probable cause of Algol's variability; though it took another 107 years before H. C. Vogel—in recognizing Algol as a spectroscopic binary whose conjunctions coincide with the moments of its light minima—confirmed this surmise in 1890 by independent observational evidence.

And more; for it was not till 97 years after Goodricke published his famous hypothesis that a first attempt was made to analyze Algol's observed light changes in terms of a rudimentary geometrical model (cf. Pickering, 1880); and that (almost) another century had to elapse before an analysis of such light changes was rewritten in terms of formalism more suitable for its interpretation (cf. Kopal, 1975a) in the frequency-domain. Such a strategy eventually led to a "break" of the code in which the information provided by eclipsing variables is being transmitted through space.

The delay in such a change of strategy was only to the detriment of a good cause (and discouragement of the observers); but it happens but seldom in the annals of science that one succeeds in hitting the nail on the head at the first trial. All mathematics necessary for our purpose was known at the time of—if not Goodricke, then certainly of William Herschel or C. F. Gauss; but, unfortunately, the underlying celestial phenomena failed to attract the attention of that Prince of the Mathematicians; and the subject continued to languish until almost our own times. Even now we are only at the beginning of the story; and if the complexity of certain tasks arising in this connection (and we shall come to grips with them in Chapters IV and VI of this book) continues to exasperate the more practically-minded readers, the latter should keep in mind that the phenomena of eclipses have never yet caused an astronomer to lose his head—at least since the days of our somewhat legendary predecessors Hi and Ho in old China! Stellar eclipses will, to be sure, have scarcely ever again an opportunity to stop a battle as (allegedly) did the famous solar eclipse of 585 B.C. at the time of Thales. However, the results of their study may well exert a more profound and lasting effect on our science than did that abortive skirmish between the Lydians and the Medes on the history of the human race.

In this spirit, the contents of this book would be incomplete if at least a mention has not been made in its concluding Chapter VII of the possibility to utilize the subjects discussed in the preceding chapters also for other than astronomical application. As is well known, one hallmark of the second half of our turbulent century—apart from space travel—has been a dramatic efflorescence of new classes of automatic computing devices which are influencing profoundly many aspects of our civilization. Although most part of this story—concerning as it does mainly electrical engineering—is of course wholly outside the scope of this book, its contents would be incomplete without at least a brief mention of a possibility that laboratory simulation of astronomical eclipse phenomena can lend itself also for the design and operation of optical analog computers which could potentially displace the digital computers in a very short time. This book is not the place in which to develop such ideas in any detail; but its last chapter will outline at least the theoretical possibilities of such a development—in the hope that its contents may stimulate some readers to translate these ideas into more practical terms.

Chapter II

LIGHT CHANGES DUE TO ECLIPSES OF SPHERICAL STARS

In the introductory chapter of this book we outlined already the full range of information—much of it unobtainable from any other source—stored in the light changes of eclipsing binary systems, which could be developed from them by appropriate analysis. The observable aspects of such messages possess many facets: from periodic variations in intensity exhibited by the systems to its coherence, polarization, spectral composition, etc. All can be recorded (by suitable instruments) to a different degree of accuracy. However, information reaching us through these diverse channels is not being sent out "*en clair*", but is encoded in a language to which we first must find a clue. In order to obtain it, in the next two chapters of this book we shall construct theoretical models of such phenomena which should serve as a "code" to the language we shall try to decipher; while in Chapters V and VI we shall expose the details of the decoding procedures which can be based upon it.

Accordingly, our opening task should then be to outline a geometrical theory of the eclipses of close binary systems which—by virtue of a chance inclination of their orbital planes to the line of sight—can become eclipsing variables. In doing so we shall regard their components as configurations of minimum potential energy—which are spheres—exhibiting circular discs on the celestial sphere normal to the line of sight. At first, so restricted a problem may seem to admit of only very restricted application to the real phenomena observed in the sky; for how could two stars situated in the proximity of each other retain spherical shape? This constitutes indeed a valid point; and the limitations of such a model will be removed in Chapter IV.

However, to come to grips with problems arising from this source in a gradual manner, in the first part of this chapter we shall investigate expressions for the loss of light due to mutual eclipses of spherical stars in terms of an elementary calculus, and shall proceed to generalize these in its second part by regarding this loss of light as a cross-correlation of two circular beams. As we shall prove, such a formulation of our problem is much more general, and better adapted for its inversion, in Chapter IV aiming to express the geometrical elements of the eclipses in terms of the observed changes of light.

II.1 Loss of Light: Eclipse Functions

In accordance with the restrictions imposed in the introductory paragraphs of this chapter, let us consider a system consisting of two stars of luminosities L_1 and L_2, revolving around a common centre of gravity in an orbit inclined to a plane tangent to the celestial sphere (at a point intersected by the line of sight) by an angle j (for its definition, cf. Eqs. (1.23) of Chapter IV). The two components—regarded as spherical—should appear in projection on the celestial sphere as circular discs; and if their dimensions are expressed in terms of their separation, we shall hereafter refer to them as fractional radii $r_{1,2}$ of the two stars.

If the sum $r_1 + r_2 < \cos j$, both components of the system would remain continuously in sight of a distant observer; and the sum $L_1 + L_2$ of their luminosities would remain constant: such a binary would emit constant light; and nothing obvious would distinguish it from the neighbouring field stars. Should, however, $r_1 + r_2 > \cos j$, the light of such a system could no longer remain constant at all times; for twice in the course of each orbital cycle one component would step out in front of its mate to *eclipse* (partly or wholly) the apparent disc of the star behind—a phenomenon whose alternation would, to a distant observer, disclose its nature by a characteristic *variation of light* of the system as a whole.

In more specific terms, the light of a binary system consisting of spherical stars—constant between eclipses—would exhibit two minima of light as the components eclipse alternately each other at the time of conjunctions. If, moreover, the orbits of the components are circular (not otherwise!), both minima will be symmetrical, of equal duration, and separated in time by exactly half the period of the orbit. At any moment during a minimum one component will eclipse a certain area (and thus cut off a certain amount of light) of the apparent disc of the other. Half a revolution later—at the corresponding phase during the other minimum—the geometrical relations of the two discs will be exactly the same, except that now the other star is in front and eclipses an equal area—though not necessarily an equal proportion—of the disc of the first star. Moreover, the deeper minimum of the two usually (though not necessarily always) corresponds to the star of greater surface brightness; though whether this star is the small or the larger of the two cannot be decided by inspection and remains to be established by subsequent investigation.

The *loss of light* $\Delta\mathcal{L}$ suffered at any phase of eclipse can general be expressed as

$$\Delta\mathcal{L} = J \cos\gamma \, d\sigma, \tag{1.1}$$

where J represents the distribution of brightness over the apparent disc of the star undergoing eclipse, of surface element $d\sigma$; and γ, the angle between the surface normal and the line of sight (i.e., one of foreshortening); while the range S of integration is to be extended over the entire eclipsed area (where $\gamma \leq 90°$).

The distribution of brightness J will, in general, be a function of the cosine of the angle of foreshortening; and can be specified from the appropriate equations of radiative transfer in stellar atmospheres. For spherical stars, it is generally legitimate to approximate their solution in the limit of zero optical depth by a finite expansion of the form

$$J = H(1 - u_1 - u_2 - \cdots - u_N + u_1 \cos \gamma + u_2 \cos^2 \gamma + \cdots + u_N \cos^N \gamma), \quad (1.2)$$

where H stands for the intensity of radiation emerging normally to the surface, and the u_n's are the "coefficients of limb-darkening" of n-th degree ($n = 1, 2, 3, \ldots N$); and, for sufficiently large N, the approximation represented by Eq. (1.2) can attain an arbitrary degree of accuracy.

Numerical values of the u_n's have been listed by Kopal (1949) for $N = 1(1)4$ in the case of plane-parallel grey atmospheres. In more general situations the coefficients u_n should, however, be regarded as additional unknowns, to be determined from an analysis of the light curve simultaneously with all other charcteristics of the respective eclipsing system.

In order to evaluate the loss of light $\Delta \mathcal{L}$ as defined by Eq. (1.1), we find it convenient to express the geometry of our problem in terms of rectangular xy-coordinates defining a plane perpendicular to the line of sight; with the origin at the centre of the disc undergoing eclipse, and the positive x-axis oriented constantly in the direction of the projected centre of the eclipsing star.

Let—in what follows and throughout this book—r_1 be the fractional radius of the star undergoing eclipse; and r_2, that of the eclipsing body (regardless of whether $r_1 \gtrless r_2$). If so, the boundary of the eclipsed disc will, in our xy-coordinates defined above, be given by the equation

$$x^2 + y^2 = r_1^2, \quad (1.3)$$

while the intersection of the shadow cylinder with the xy-plane will be given by

$$(\delta - x)^2 + y^2 = r_2^2, \quad (1.4)$$

where

$$\delta^2 = \sin^2 \psi \sin^2 j + \cos^2 j = 1 - \cos^2 \psi \sin^2 j \quad (1.5)$$

stands for the apparent (projected) separation of the centres of the two discs (see Figure II.1) expressed in terms of their maximum separation taken as the unit, and ψ, the phase angle (i.e., the true anomaly of the relative orbit reckoned from the moment of conjunction, identified with the primary minimum), as defined by Eqs. (1.23) of Chapter IV.

Outside eclipses, the luminosity L_1 of the component of fractional radius r_1—limb-darkened to the N-th degree in accordance with Eq. (1.2), will be given by

$$L_1 = 2\pi \int_0^{r_1} J \sin^{-1}(r/r_1) r \, dr =$$

$$= \pi r_1^2 H \left\{ 1 - \sum_{n=0}^{N} \frac{n \, u_n}{n + 2} \right\}; \quad (1.6)$$

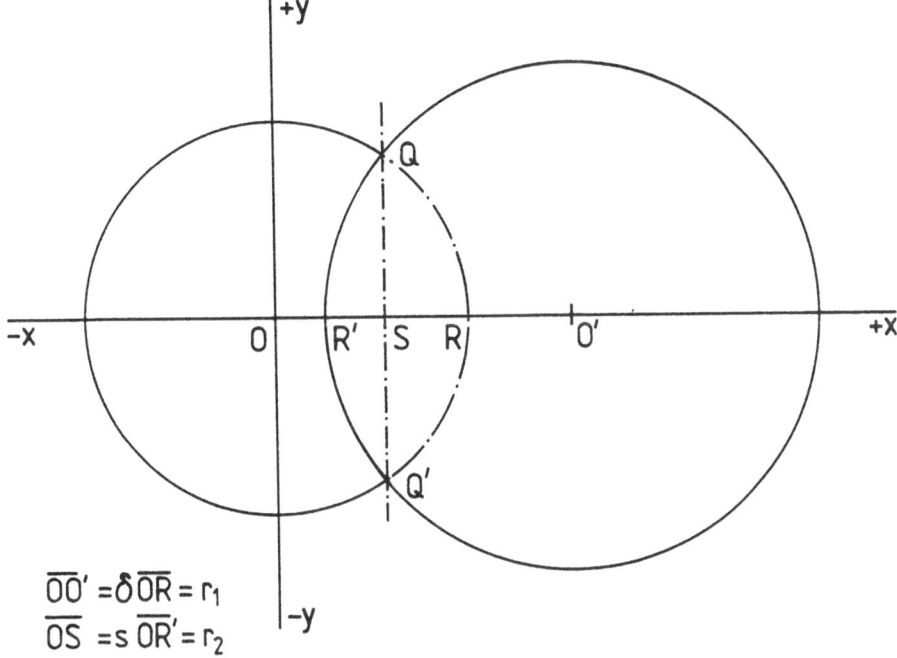

$$\overline{OO'} = \delta \, \overline{OR} = r_1$$
$$\overline{OS} = s \, \overline{OR'} = r_2$$

Figure II.1: Geometry of the eclipses of spherical stars

but to establish the fractional loss of light $\delta \mathcal{L}/L_1$ within eclipses we must first specify the limits S of integration on the r.h.s. of Eq. (1.1).

Provided that $\delta < r_1 + r_2$, the circles defined by Eqs. (1.3) and (1.4) will intersect (cf. Figure II.1) at points $Q(s, \pm\sqrt{r_1^2 - s^2})$, with

$$s = \frac{r_1^2 - r_2^2 + \delta^2}{2\delta} \qquad (1.7)$$

constrained to vary between $+r_1$ at the moment of first contact of the eclipse to $-r_1$ at the commencement of totality. Since, furthermore,

$$r_1 \cos \gamma = \sqrt{r_1^2 - x^2 - y^2} \qquad (1.8)$$

and

$$\cos \gamma \, d\sigma = dx \, dy , \qquad (1.9)$$

the fractional loss of light during the eclipse of a star whose surface brightness falls off from centre to limb in accordance with a law of N-th degree of the form (1.2) can be expressed as

$$\alpha \equiv \frac{\delta \mathcal{L}_1}{L_1} = \sum_{n=0}^{N} C^{(n)} \alpha_n^0 , \qquad (1.10)$$

where

$$C^{(0)} = \frac{1 - u_1 - u_2 - \cdots u_N}{1 - \sum_{n=1}^{N} \frac{n u_n}{n+2}} ; \tag{1.11}$$

while, for $n > 0$,

$$C^{(n)} = \frac{u_n}{1 - \sum_{n=1}^{N} \frac{n u_n}{n+2}} ; \tag{1.12}$$

and

$$\pi r_1^{n+2} \alpha_n^0 = \left\{ \int_s^{r_1} \int_{-\sqrt{r_1^2 - x^2}}^{\sqrt{r_1^2 - x^2}} + \int_{\delta - r_2}^{s} \int_{-\sqrt{r_2^2 - (\delta - x)^2}}^{\sqrt{r_2^2 - (\delta - x)^2}} \right\} z^n \, dx \, dy , \tag{1.13p}$$

where, in accordance with Eq. (1.8), $z \equiv r_1 \cos \gamma$.

Equation (1.13p) holds good, to be sure, only throughout *partial* phases of the eclipses—i.e., if

$$r_1 + r_2 > \delta > |r_1 - r_2| \tag{1.14}$$

irrespective of whether the eclipsed star of fractional radius r_1 is the larger or smaller of the two. In this latter case—when $r_1 < r_2$—we shall hereafter refer (for obvious analogy) to such an eclipse as an *occultation*; whereas if the converse is true (i.e., if $r_1 > r_2$) we shall term it a *transit*. If, moreover, $\delta < r_2 - r_1$ in the course of an occultation, the eclipse of the star of radius r_1 becomes *total*—in which case Eq. (1.13p) reduces to a constant equal to

$$\alpha_n^0 = \frac{2}{n+2} \tag{1.15}$$

during totality; while whenever $\delta < r_1 - r_2$ during a transit eclipse, the latter becomes *annular* and Eq. (1.13p) should then be replaced by

$$\pi r_1^{n+2} \alpha_n^0 = \int_{\delta - r_2}^{\delta + r_2} \int_{-\sqrt{r_2^2 - (\delta - x)^2}}^{\sqrt{r_2^2 - (\delta - x)^2}} z^n \, dx \, dy . \tag{1.13a}$$

The above-defined functions α_n^0—hereafter referred to as "associated α-functions of index zero and order n"—play a fundamental role in the mathematical theory of stellar eclipses and much of the contents of this book will revolve around their use. In order to prepare them more fully for this role, let us generalize them for an arbitrary (integral) index m by setting

$$\pi r_1^{m+n+2} \alpha_n^m = A_n^m + B_n^m , \tag{1.16}$$

where, for partial eclipses (occultations or transits),

$$A_n^m = \int_s^{r_1} \int_{-\sqrt{r_1^2 - x^2}}^{\sqrt{r_1^2 - x^2}} x^m z^n \, dx \, dy , \tag{1.17}$$

and

$$B_n^m = \int_{\delta - r_2}^{s} \int_{-\sqrt{r_2^2 - (\delta - x)^2}}^{\sqrt{r_2^2 - (\delta - x)^2}} x^m z^n \, dx \, dy . \tag{1.18}$$

Should the eclipse become annular, $\mathcal{A}_n^m = 0$ and $\mathcal{B}_n^m$ continues to be given by Eq. (1.18) provided, however, that the quantity s in the upper limit of the first integral on the r.h.s. of (1.18) has been replaced by $\delta + r_2$.

As to the expression (1.17) for $\mathcal{A}_n^m$, integrating with respect to y we find at once that

$$
\mathcal{A}_n^m = B\left(\frac{1}{2}, \frac{n+2}{2}\right) \int_s^{r_1} x^m (r_1^2 - x^2)^{(n+1)/2} dx =
$$
$$
= B\left(\frac{1}{2}, \frac{n+2}{2}\right) \{D_{n+1}^m(r_1) - D_{n+1}^m(s)\}, \tag{1.19}
$$

where B denotes a complete beta-function, and

$$
D_{n+1}^m(x) = \int_0^x x^m (r_1^2 - x^2)^{(n+1)/2} dx \tag{1.20}
$$

represents a binomial integral tractable by elementary means. In point of fact,

$$
D_{n+1}^m(r_1) = \frac{1}{2} B\left(\frac{m+1}{2}, \frac{n+3}{2}\right) r_1^{m+n+2} \tag{1.21}
$$

and

$$
D_{n+1}^m(s) = \frac{1}{m+1} \left(\frac{s}{r_1}\right)^{m+1} {}_2F_1\left(-\frac{n+1}{2}, \frac{m+1}{2}, \frac{m+3}{2}; \frac{s^2}{r_1^2}\right) r_1^{m+n+2}; \tag{1.22}
$$

where, for odd values of n, the hypergeometric series on the right-hand side reduces to a polynomial. In point of fact, for any value of n, Eq. (1.17) can be expressed as

$$
\mathcal{A}_n^m = \pi B\left(\frac{1}{2}, \frac{n+2}{2}\right) r_1^{m+n+2} J_{n+1,0}^m(\kappa) \tag{1.23}
$$

in terms of the integrals $J_{\beta,\gamma}^m$ defined by Eq. (2.24) in the next section of this chapter, with the modulus

$$
\kappa^2 = \frac{1}{2}\left(1 - \frac{\delta - s}{r_2}\right) = \frac{r_2^2 - (\delta - r_2)^2}{4\delta r_2}, \tag{1.24}
$$

established by means of the identity $r_1^2 - s^2 = r_2^2 - (\delta - s)^2$; and would by itself constitute the entire loss of light if the eclipsing component would act as a straight occulting edge (see p.54).

The evaluation of the function $\mathcal{B}_n^m$ as defined by Eq. (1.18) above proves to be somewhat more complicated. Integrating (1.18) with respect to y we find that, if $n \equiv 2\nu(\nu = 1, 2, \ldots)$ is zero or an even integer,

$$
\mathcal{B}_{2\nu}^m = \frac{1}{\pi} B\left(\frac{1}{2}, 1+\nu\right) \sum_{j=0}^{\nu} B\left(\frac{1}{2}, \frac{1}{2}+\nu-j\right) I_{2j,1,2(\nu-j)}^m; \tag{1.25}
$$

while if $n \equiv 2\nu - 1$ is odd,

$$B_{2\nu-1}^m = \frac{1}{\pi} B\left(\frac{1}{2}, \frac{1}{2} + \nu\right) \left\{\sum_{j=1}^{\nu} B\left(\frac{1}{2}, j\right) I_{2(\nu-j),1,2j-1}^m + 2\Pi_{2\nu}^m\right\}, \qquad (1.26)$$

where

$$I_{2\alpha,\beta,\gamma}^m = \int_{\delta-r_2}^{c_2} (r_1^2 - x^2)^\alpha [r_2^2 - (\delta - x^2)]^{\beta/2} [2\delta(s - x)]^{\gamma/2} x^m \, dx, \qquad (1.27)$$

and

$$\Pi_{2\nu}^m = \int_{\delta-r_2}^{c_2} x^m (r_1^2 - x^2)^\nu \sin^{-1} \sqrt{\frac{r_2^2 - (\delta - x)^2}{r_1^2 - x^2}} \, dx ; \qquad (1.28)$$

where

$$c_2 = s \text{ or } \delta + r_2 \qquad (1.29)$$

depending on whether the eclipse is partial or annular; and in which advantage has been taken of the fact that the equations

$$x^2 + y^2 + z^2 = r_1^2 \qquad (1.30)$$

of the surface of the star undergoing eclipse and of the shadow cylinder cast by its companion, as given by (1.4), can be solved for z in terms of x yielding

$$z^2 = 2\delta(s - x), \qquad (1.31)$$

where s continues to be given by (1.7).

The method of evaluation of the integral on the right-hand side of Eq. (1.25) depends, in principle, upon the values of the subscripts α, β and γ as well as of the superscript m on its left-hand side. As $x < r_1$, the first can be easily suppressed by an appeal to the binomial theorem, yielding

$$I_{2\alpha,\beta,\gamma}^m = \sum_{i=0}^{\alpha} (-1)^i \binom{\alpha}{i} r_1^{2(\alpha-i)} I_{0,\beta,\gamma}^{m+2i} \qquad (1.32)$$

and

$$I_{0,\beta,\gamma}^{m+2i} = \pi r_2^{\beta+\gamma+1} \sum_{j=0}^{m+2i} (-1)^j \binom{m+2i}{j} \delta^{m+2i} (r_2/\delta)^j I_{\beta,\gamma}^j, \qquad (1.33)$$

where we have abbreviated

$$\pi r_2^{\beta+\gamma+j+1} I_{\beta,\gamma}^j = \int_{\delta-r_2}^{c_2} [r_2^2 - (\delta - x)^2]^{\beta/2} [2\delta(s - x)]^{\gamma/2} (\delta - x)^j dx . \qquad (1.34)$$

Like the α_n^m's, the $I_{\beta,\gamma}^j$'s so defined stand for another important family of auxiliary functions, which we shall frequently encounter in the first part of this book. The nature of problems to be investigated with their aid is such that (on account of the symmetry of the configuration, within eclipses, shown on Figure

II.1, with respect to the y-axis) the parameter β can assume the values of odd integers only; and the finiteness of the integrals on the right-hand side of (1.34) requires that the index j—i.e., m—must be non-negative (i.e., zero or a positive) integer. The form of the functions α_n^m as well as of $I_{\beta,\gamma}^m$ depends, therefore, on the remaining parameter γ.

The form (1.2) of the adopted law of limb-darkening restricts again γ to be an integer; and the finiteness of both α_n^m as well as $I_{\beta,\gamma}^m$ requires that $\gamma \geq -1$. For γ zero or an even integer, both functions defined by Equations (1.27) and (1.28) or (1.34) are elementary, and can be expressed in terms of circular and algebraic functions (cf. Kopal, 1942b); but if γ happens to be odd, the integral on the right-hand side of Eq. (1.34) becomes elliptic.

In order to evaluate it (cf. again Kopal, 1942b) we find it convenient to change over from x to a new variable (say) t, defined so that

$$[r_2^2 - (\delta - x)^2][s - x] = (t - e_1)(t - e_2)(t - e_3) \tag{1.35}$$

subject to the conditions that

$$e_1 > e_2 > e_3 \quad \text{and} \quad e_1 + e_2 + e_3 = 0. \tag{1.36}$$

Obviously

$$\left.\begin{array}{rcl} e_1 &=& +\frac{1}{3}(\delta - s) + r_2 \\ e_2 &=& -\frac{2}{3}(\delta - s) \\ e_3 &=& +\frac{1}{3}(\delta - s) - r_2 \end{array}\right\} \tag{1.37p}$$

if the eclipse is partial, and

$$\left.\begin{array}{rcl} e_1 &=& -\frac{2}{3}(\delta - s) \\ e_2 &=& +\frac{1}{3}(\delta - s) + r_2 \\ e_3 &=& +\frac{1}{3}(\delta - s) - r_2 \end{array}\right\} \tag{1.37a}$$

if it is annular. In either case we are, therefore, entitled to set

$$t \equiv \wp(v), \tag{1.38}$$

where $\wp$ denotes the Weierstrass elliptic π-functions of argument v, replacing t as our independent variable.

The integrals on the right-hand side of Equations (1.33) in terms of this new variable become

$$I_{0,\beta,\gamma}^m = -2i^{\beta+\gamma}(2\delta)^{\gamma/2} \int_{\omega_2}^{\omega_1+\omega_2} \{[\wp(v) - e_1] \times$$

$$\times[\wp(v) - e_3]\}^{(\beta+1)/2} \{\wp(v) - e_2\}^{\gamma+1)/2}\{\wp(v) + h\}^m dv, \tag{1.39}$$

where

$$h \equiv \sum_{j=1}^{3} e_j = \frac{1}{3}(2\delta + s) \tag{1.40}$$

and $i \equiv \sqrt{-1}$ stands for the imaginary unit; with limits $\omega_{1,2}$ defined by

$$\wp(\omega_1) = e_1 \text{ and } \wp(\omega_2) = e_3 . \tag{1.41}$$

If, moreover, $\beta = 1$ (as is the case in Eqs. 1.25 and 1.26 for B_n^m), Equation (1.39) can undergo further reduction. For if so, Eq. (1.39) can obviously be rewritten as

$$I_{0,1,\gamma}^m = -2i^{\gamma+1}(2\delta)^{\gamma/2} \int_{\omega_2}^{\omega_1+\omega_2} [\wp(v) - e_1][\wp(v) - e_2] \times$$
$$\times [\wp(v) - e_3][\wp(v) - e_2]^{(\gamma-1)/2}[\wp(v) + h]dv =$$

$$= -\frac{i^{\gamma+1}}{2}(2\delta)^{\gamma/2} \int_{\omega_2}^{\omega_1+\omega_2} [\wp(v) - e_2]^{(\gamma-1)/2}[\wp(u) + h]^m[\wp'(v)]^2 dv, \tag{1.42}$$

where the accent on $\wp(v)$ denotes a derivative with respect to v. Indeed, the Weierstrass π-function is known to satisfy the first-order quadratic equation of the form

$$\{\wp'(v)\}^2 = 4[\wp(v) - e_1][\wp(v) - e_2][\wp(v) - e_3] =$$

$$= 4\wp^3(v) - g_2\wp(v) - g_3 , \tag{1.43}$$

where

$$g_2 = -4(e_1e_2 + e_1e_3 + e_2e_3) \text{ and } g_3 = 4e_1e_2e_3 . \tag{1.44}$$

Moreover, γ being an odd integer, the entire integrand in (1.42) can be compressed as a polynomial in ascending powers of $\wp(v)$; and all that remains to be done is to reduce the integrals

$$\int_{\omega_2}^{\omega_1+\omega_2} \{\wp(v)\}^j dv \text{ for } j = 0, 1, 2, \ldots, \frac{1}{2}(\gamma - 1) + m + 3 . \tag{1.45}$$

to the Legendre normal forms. This can, in turn, be done by expressing (by successive differentiation of (1.43)) the j-th powers of $\wp(v)$ in terms of its derivatives. In doing so (and remembering that odd derivatives of $\wp(v)$ with arguments ω_2 as well as $\omega_1 + \omega_2$ vanish), we readily see that

$$\int_{\omega_2}^{\omega_1+\omega_2} dv = \omega_1 = \frac{F(\frac{\pi}{2}, \kappa)}{\sqrt{e_1 - e_3}} , \tag{1.46}$$

$$\int_{\omega_2}^{\omega_1+\omega_2} \wp(v)dv \equiv \eta_1 = \sqrt{e_1 - e_3}\, E\left(\frac{\pi}{2}, \kappa\right) - \frac{e_1}{\sqrt{e_1 - e_3}} F\left(\frac{\pi}{2}, \kappa\right) , \tag{1.47}$$

$$\int_{\omega_2}^{\omega_1+\omega_2} \{\wp(v)\}^2 dv = \frac{1}{12} g_2\omega_1 , \tag{1.48}$$

etc.; where F and E stand for the Legendre complete elliptic integral of the first and second kind, with the modulus

$$\kappa^2 = \frac{e_2 - e_3}{e_1 - e_3}, \qquad (1.49)$$

identical with that introduced by Eq. (1.24). The reader may note from Equations (1.36) that the moduli appropriate for partial and annular eclipses are mutually reciprocal.

After having thus established the explicit form of Equations (1.13) in a finite number of terms for any value of the subscripts and of m, let us return to Equation (1.28). Integrating Π by parts we obtain

$$\Pi_{2\nu}^m = GD_{2\nu}^m(s) - \sqrt{\frac{\delta}{2}} \int_{c_1}^{c_2} \frac{X D_{2\nu}^m(x)dx}{\sqrt{(x - c_1)(x - c_2)(x - c_3)}}, \qquad (1.50)$$

where $G = \frac{1}{2}\pi$ or 0 depending on whether the eclipse is partial or annular,

$$X \equiv \frac{x^2 - 2sx + r_1^2}{r_1^2 - x^2}, \qquad (1.51)$$

and

$$c_1 = \delta + r_2, \quad c_2 = s, \quad c_3 = \delta - r_2, \qquad (1.52\text{p})$$

if the eclipse is partial; and

$$c_1 = s, \quad c_2 = \delta + r_2, \quad c_3 = \delta - r_2, \qquad (1.52\text{a})$$

if it is annular. Substitute, as before,

$$x - \frac{1}{3}\sum_{j=1}^{3} c_j = \wp(u) \qquad (1.53)$$

and expand

$$X D_{2\nu}^m(x) = \sum_{j=0}^{m+n+2} a_n^m(j)\{\wp(u)\}^j + b_n^m r_1^{m+n+2} \times$$

$$\times \left\{ \frac{r_1 - s}{\wp(u) + h - r_1} + (-1)^m \frac{r_1 + s}{\wp(u) + h + r_1} \right\}, \qquad (1.54)$$

where the coefficients a_n^m are polynomials of the $(m + n + 2 - j)$th degree in r_1, δ, and s; and b_n^m is a positive fraction (numerical factor). Since, by definition,

$$dx = 2\sqrt{(x - c_1)(x - c_2)(x - c_3)}du, \qquad (1.55)$$

we see that the $\Pi_{2\nu}^m$ can be expressed in terms of integrals of powers of $\wp(u)$, which we have just solved, plus two integrals of the form

$$\int_{\omega_2}^{\omega_1 + \omega_2} \frac{du}{\wp(u) + h \pm r_1},$$

which are new and remain to be evaluated.

In order to do so, we introduce new arguments $v_{1,2}$ defined by

$$- (h \pm r_1) \;=\; \wp(v_{1,2}) \,. \tag{1.56}$$

Then, by means of a well-known theorem [1] we have

$$\int_{\omega_1}^{\omega_1+\omega_2} \frac{du}{\wp(u)-\wp(v_j)} \;=\; \frac{2}{\wp'(v_j)}\{\omega_1\zeta(v_j)-\eta_1 v_j\}, \quad j=1,2\,, \tag{1.57}$$

where the accent denotes a derivative with respect to v_1 and ζ is the Weierstrass zeta-function. As one can easily verify,

$$\wp'(v_{1,2}) \;=\; \mp 2i\sqrt{2\delta}(r_1 \pm s) \tag{1.58p}$$

if the eclipse is partial, and

$$\wp'(v_{1,2}) \;=\; -2i\sqrt{2\delta}(r_1 \pm s) \tag{1.58a}$$

if it is annular. In order to remove the imaginary unit we put

$$v_1 \;=\; iw_1, \quad v_2 \;=\; iw_2 + \omega_1 \,. \tag{1.59}$$

If we remember that

$$\zeta(\omega_1) \;=\; \eta_1, \quad \wp(\omega_1) \;=\; e_1, \quad \wp'(\omega_1) \;=\; 0 \,, \tag{1.60}$$

the addition theorem for Weierstrass zeta-functions yields

$$\zeta(\omega_1 + iw_j) \;=\; \eta_1 + \zeta(iw_j) + \frac{i}{\sqrt{2\delta}}\{r_1 - (\delta - r_2)\} \tag{1.61p}$$

if the eclipse is partial, and

$$\zeta(\omega_1 + iw_j) \;=\; \eta_1 + \zeta(iw_j) + i\sqrt{2\delta} \tag{1.61a}$$

if it is annular. If we further substitute

$$\zeta(iw) \;=\; -i\zeta^*(w) \,, \tag{1.62}$$

where

$$\zeta^*(w; e_1, e_2, e_3) \;=\; \zeta(w; -e_1, -e_2, -e_3) \,, \tag{1.63}$$

we finally obtain that, for partial eclipses,

$$\sqrt{2\delta}(r_1 + s) \int_{\omega_2}^{\omega_1+\omega_2} \frac{du}{\wp(u)-\wp(v_1)} \;=\; \omega_1\zeta^*(w_1) + \eta_1 w_1 \,, \tag{1.641}$$

---

[1] Cf. Whittaker and Watson (1920), Section 20.53.

and

$$\sqrt{2\delta}(r_1 - s) \int_{\omega_2}^{\omega_1 + \omega_2} \frac{du}{\wp(u) - \wp(v_2)} = -\omega_1 \zeta^*(w_2) - \eta_1 w_2 + \frac{\omega_1}{\sqrt{2\delta}}(r_1 + r_2 - \delta). \tag{1.642}$$

If, however, the eclipse is annular, Equation (1.641) continues to hold good; but Equation (1.642) is to be replaced by

$$\sqrt{2\delta}(r_1 - s) \int_{\omega_2}^{\omega_1 + \omega_2} \frac{du}{\wp(u) - \wp(v_2)} = \omega_1 \zeta^*(w_2) + \eta_1 w_2 - \omega_1 \sqrt{2\delta}. \tag{1.65}$$

The functions $w_{1,2}$ and $\zeta^*(w_{1,2})$, expressed in terms of Legendre normal forms, become

$$w_{1,2} = \frac{F(\phi_{1,2}, \kappa')}{\sqrt{e_1 - e_3}} \tag{1.66}$$

and

$$\zeta^*(w_{1,2}) = e_3 w_{1,2} + \sqrt{e_1 - e_3} E(\phi_{1,2}, \kappa') + \frac{1}{\sqrt{2\delta}}[r_1 \pm (\delta - r_2)], \tag{1.67p}$$

$$\zeta^*(w_{1,2}) = \dot{e}_3 w_{1,2} + \sqrt{e_1 - e_3} E(\phi_{1,2}, \kappa') + \sqrt{2\delta}, \tag{1.67a}$$

—as to whether the eclipse is partial (p) or annular (a)—where the complementary modulus κ is defined by

$$(\kappa')^2 = \frac{e_1 - e_2}{e_1 - e_3} = 1 - \kappa^2, \tag{1.68}$$

and the amplitudes for partial and annular eclipses take the respective forms

$$\phi_1 = \sin^{-1}\sqrt{\frac{2r_2}{r_1 + r_2 + \delta}}, \quad \phi_2 = \sin^{-1}\sqrt{\frac{2\delta}{r_1 + r_2 + \delta}}, \tag{1.69p}$$

and

$$\phi_1 = \phi_2 = \sin^{-1}\sqrt{\frac{r_1 + r_2 - \delta}{r_1 + r_2 + \delta}}. \tag{1.69a}$$

Let us put, for brevity's sake,

$$(r_1 + s) \int_{\omega_2}^{\omega_1 + \omega_2} \frac{du}{\wp(u) - \wp(v_1)} \mp (r_1 - \delta) \int_{\omega_3}^{\omega_1 + \omega_2} \frac{du}{\wp(u) - \wp(v_2)} \equiv \frac{1}{\sqrt{2\delta}} \mathcal{E}_{1,2}. \tag{1.70}$$

By combination of the above formulae it follows that, for partial eclipses,

$$\mathcal{E}_{1,2} = \left\{ E\left(\frac{\pi}{2}, \kappa\right) - F\left(\frac{\pi}{2}, \kappa\right) \right\} \{F(\phi_1, \kappa') \pm F(\phi_2, \kappa')\} +$$
$$+ F\left(\frac{\pi}{2}, \kappa\right) \{E(\phi_1, \kappa') \pm E(\phi_2, \kappa) + \kappa \cos \phi_1 \sec \phi_2\}. \tag{1.71}$$

If the upper sign is valid, this expression admits of a drastic simplification; for, by an obvious extension of a theorem due to Legendre, [2] the reader should have no difficulty to prove that

$$\mathcal{E}_1 = \frac{\pi}{2} + \sqrt{\frac{\delta}{r_2}} F\left(\frac{\pi}{2}, \kappa\right) . \tag{1.72}$$

Hence, by subtraction of $\mathcal{E}_1$ and $\mathcal{E}_2$, the latter takes the form

$$\mathcal{E}_2 = \frac{\pi}{2} - 2\left\{ E\left(\frac{\pi}{2}, \kappa\right) - F\left(\frac{\pi}{2}, \kappa\right) \right\} F(\phi_2, \kappa') -$$

$$- 2\left\{ E(\phi_2, \kappa') - \frac{1}{2}\sqrt{\frac{\delta}{r_2}} \right\} F\left(\frac{\pi}{2}, \kappa\right), \tag{1.73}$$

in which both kinds of incomplete integrals possess a common amplitude.

If, finally, the eclipse is annular we similarly obtain

$$\mathcal{E}_1 = \kappa\sqrt{\frac{\delta}{r_2}} F\left(\frac{\pi}{2}, \kappa\right) \tag{1.74}$$

and

$$\mathcal{E}_2 = 2F(\phi_2, \kappa')\left\{ E\left(\frac{\pi}{2}, \kappa\right) - F\left(\frac{\pi}{2}, \kappa\right) \right\} +$$

$$+ 2F\left(\frac{\pi}{2}, \kappa\right)\left\{ E(\phi_2, \kappa') - \frac{\kappa}{2}\sqrt{\frac{\delta}{r_2}} \right\} . \tag{1.75}$$

This completes the evalution of the "circular integrals" α_n^m associated with the effects of *tidal* distortion. We found that the expressions for A_n^m are all elementary (in fact, polynomial if n is odd); while those for B_n^m are such only if n is zero or an even integer: if n were odd, the corresponding B_n^m's can be evaluated exactly only in terms of elliptic integrals. Expressions of the form $I_{0,1,\gamma}^m$ where γ is a positive odd integer, or (if m is also odd) the $\Pi_{2\nu}^m$'s, can likewise be expressed in terms of Legendre complete integrals of the first and second kind. If, however, m is zero or even, the $\Pi_{2\nu}^m$'s are bound to involve also complete elliptic integrals of the third kind which belong to the "circular" class and are, therefore, expressible in terms of incomplete integrals of the first and second kind with complementary moduli.

The analytical expressions for the associated α-functions as well as of the I-integrals introduced in this section hold good at any phase of the eclipses— including the end points; and their explicit forms for integral values of $m = 0(1)4$ and $n = -1(1)4$ have been evaluated by the author of this book (cf. Kopal,

---

[2] Cf. again Whittaker and Watson (1920), section 22.735

1942b) in a closed form almost fifty years ago. Numerically, they all prove to be quantities of zero order; and such that

$$\alpha_{n+1}^m < \alpha_n^m \quad \text{and} \quad \alpha_n^{m+1} \ll \alpha_n^m \,. \tag{1.76}$$

They all vanish at the moments of the first (and fourth contact)—i.e., when $\delta = r_1 + r_2$—and, for odd values of m, also at the second contact of occultation eclipses (when $\delta = r_2 - r_1$). If, however, m happens to be even ($= 2\mu$) and the eclipse is an *occultation* whose inner contact marks the commencement of totality, then

$$\alpha_{2\nu}^{2\mu} = \frac{\nu! \, \Gamma(\mu + \frac{1}{2})}{\sqrt{\pi} \, (\mu + \nu + 1)!} \tag{1.77}$$

if the order $n \equiv 2\nu$ is even, and

$$\alpha_{2\nu-1}^{2\mu} = \frac{\Gamma(\mu + \frac{1}{2})\Gamma(\nu + \frac{1}{2})}{\sqrt{\pi}\Gamma(\mu + \nu + \frac{3}{2})} \tag{1.78}$$

if $n \equiv 2\nu - 1$ is odd. For $\mu = 0$, both (1.77) and (1.78) reduce to the value already given by Eq. (1.15).

If, on the other hand, the eclipse is a *transit* (i.e., $r_1 > r_2$) and the inner contact at which $\delta = r_1 - r_2$ marks the beginning of the annular phase, the α_n^m's remain functions of the ratio of the radii $k \equiv r_2/r_1$ regardles of whether n is even or odd. Their explicit values for arbitrary m's have not yet been established. However, if $m = 0$, it can be shown (cf. Kopal, 1975c) that, at the moment of internal tangency,

$$\alpha_n^0(k) = \frac{(n+2)! \, k^{(n+4)/2}}{\Gamma(2 + \frac{n}{2})\Gamma(3 + \frac{n}{2})} \, {}_2F_1\left(-\frac{n}{2}, 2 + \frac{n}{2}, 3 + \frac{n}{2}; k\right). \tag{1.79}$$

For even values of n, the hypergeometric series on the right-hand side of the preceding equation reduces to polynomials; and although for odd values of n this is no longer the case, the respective series can still be expressed in a closed form in terms of more elementary functions (see Appendix I of Kopal, 1979).

For $m = 0$, the eclipse functions α_n^0 as well as $I_{\beta,\gamma}^0$ during transit phase continue to vary throughout annular eclipses until—at the moment when the latter become *central* (i.e., $\delta = 0$) they reduce to

$$\alpha_n^0 = \frac{2}{n+2}\left\{1 - (1 - k^2)^{(n+2)/2}\right\} =$$
$$= k^2 \, {}_2F_1\left(-\frac{n}{2}, 1; 2; k^2\right), \tag{1.80}$$

while

$$\pi I_{\beta,\gamma}^0 = B\left(\frac{1}{2}, 1 + \frac{1}{2}\beta\right) k^{-\gamma}(1 - k^2)^{\gamma/2} \,; \tag{1.81}$$

where $k < 1$ continues to stand for the ratio r_2/r_1.

II.2   Recursion Properties of Eclipse Functions

In the preceding section of this chapter we found it possible to express the light changes due to mutual eclipses of spherical stars in terms of the associated α-functions and certain related integrals varying with the phase, and outlined the way in which such functions can be evaluated in a closed form. However, their explicit forms (cf. Kopal, 1942b) are anything but simple; in fact, they prove to be so complicated as to be of but limited use for application to practical cases: only for even values of n on the r.h.s. of the law (1.2) of limb-darkening they turned out to be expressible in terms of elementary (i.e., inverse circular and algebraic) functions; while for odd terms they require also elliptic integrals (both complete and incomplete) for their representation.

In order to bypass the difficulties arising from this source, certain general properties common to all eclipse functions will be investigated; and the best way to do so is to turn our attention to their recursion and differential properties.

In order to investigate such of their properties, we find it convenient to re-define the α_n^m-functions of different orders and indices in terms of plane-polar coordinates, of the form

$$\pi r_1^{m+n+2}\alpha_n^m \; = \; 2\int_{\delta-r_2}^{r_1}\int_0^{\theta_0} r^{m+1}(r_1^2-r^2)^{n/2}\cos^m\theta\,dr\,d\theta \, , \qquad (2.1)$$

with the upper limit θ_0 of angular integration given by the equation

$$\cos\theta_0 \; = \; \frac{\delta^2+r^2-r_2^2}{2\delta r} \, . \qquad (2.2)$$

This definition remains valid only during partial eclipses as long as $\delta \geq r_2$. When the converse becomes the case, and the eclipsing limb advances past the centre of the apparent disc of the star of fractional radius r_1, the limits of integration on the r.h.s. of Eq. (2.1) must be adjusted so that

$$\pi r_1^{m+n+2}\alpha_n^m \; = \; \left\{\int_0^{r_1}\int_0^{\pi} - \int_{\delta-r_2}^{r_1}\int_0^{\pi-\theta_0}\right\} r^{m+1}(r_1^2-r^2)^{n/2}\cos^m\theta\,dr\,d\theta \; ; \quad (2.3)$$

moreover, during annular phases of transit eclipses (when $\delta < r_1 - r_2$) the upper limits r_1 and $\pi - \theta_0$ in the second pair of integrals on the right-hand side of the foregoing equation should be replaced by $\delta + r_2$ and π, respectively.

Integrating the right-hand sides of Eqs. (2.1) and (2.3) with respect to the angular variable θ we find that

$$\int_0^{\theta_0}\cos^m\theta\,d\theta \; = \; \pi I_{-1,0}^m(\cos\theta_0) \, , \qquad (2.4)$$

where the $I_{-1,2}^m$'s are particular cases of the family of integrals defined in rectangular coordinates by Equation (1.33). Therefore, for $\delta \geq r_2$ Eq. (2.1) can be rewritten as

$$r_1^{m+n+2}\alpha_n^m \; = \; 2\int_{\delta-r_2}^{r_1} (r_1^2-r^2)^{n/2}I_{-1,0}^m(\cos\theta_0)r^{m+1}\,dr \qquad (2.5)$$

for any value of m or n. If $\delta < r_2$—and the relevant form for α_n^m is that given by Eq. (2.3)—the fact that, for $\theta_0 = \pi$,

$$I_{-1,0}^{2\mu+1}(-1) = 0 \tag{2.6}$$

if $m \equiv 2\mu + 1$ is odd, and

$$I_{-1,0}^{2\mu}(-1) = \binom{\mu - \frac{1}{2}}{\mu} \tag{2.7}$$

if $m \equiv 2\mu > 0$ is even, Eq. (2.5) can be expressed as

$$r_1^{2(\mu+1)+n+1}\alpha_n^{2\mu+1} = 2\int_{r_2-\delta}^{r_1}(r_1^2 - r^2)^{n/2}I_{-1,0}^{2\mu+1}(\cos\theta_0)r^{2(\mu+1)}dr \tag{2.8}$$

for odd values of $m(\mu = 0, 1, 2, \ldots)$; while if m is even,

$$
\begin{aligned}
r_1^{2\mu+n+2}\alpha_n^{2\mu} &= \frac{\Gamma(\mu+\frac{1}{2})\Gamma(1+\frac{1}{2}n)}{\sqrt{\pi}\,\Gamma(\mu+\frac{1}{2}n+2)}r_1^{2\mu+n+2} - \\
&\quad - 2\int_{r_2-\delta}^{r_1}(r_1^2 - r^2)^{n/2}I_{-1,0}^{2\mu}(-\cos\theta_0)r^{2\mu+1}dr. \tag{2.9}
\end{aligned}
$$

The explicit forms of the functions $I_{-1,0}^m(x)$ for any value of $m \geq 0$ are well known (cf., e.g., Eqs. (1)–(6) or Appendix IV; Kopal, 1979) to satisfy the simple recursion formula

$$mI_{-1,0}^m(x) = (m-1)I_{-1,0}^{m-2}(x) + x^{m-1}I_{-1,0}^1(x). \tag{2.10}$$

A combination of (2.10) with Eqs. (2.1)–(2.3) for the associated α-functions immediately discloses the latter to obey the recursion formula

$$(m+1)\alpha_n^{m+1} + m(\alpha_{n+2}^{m-1} - \alpha_n^{m-1}) = (2\delta/\pi r_1^{m+n+3})I_{0,1,n}^m, \tag{2.11}$$

valid for $m = 0, 1, 2, \ldots$ and $n = -1, 0, 1, 2, \ldots$; where $I_{0,1,n}^m$ stands for the respective member of the family of integrals defined by Eq. (1.26) for $\alpha = 0$, $\beta = 1$ and $\gamma = n$. With suitable limits c_2 imposed on the latter, the foregoing recursion formula (2.11) holds good for any type of eclipse—be it occultation or transit; partial, total or annular.

Eq. (2.11) does not represent the only recursion formula satisfied by the associated α-functions. Another can be constructed if we return to Eq. (2.1) and integrate its right-hand side by parts, obtaining

$$r_1^{m+n+2}\alpha_n^m = \frac{2}{n+2}\int_{\delta-r_2}^{r_1}(r_1^2 - r^2)^{(n+4)/2}\frac{d}{dr}\left\{r^mI_{-1,0}^m(\cos\theta_0)\right\}dr. \tag{2.12}$$

Replace now, in this equation, n by $n-2$ and subtract from (2.12) as it stands; in doing so we find that

$$
\begin{aligned}
r_1^{m+n+2}\{n\alpha_{n-2}^m - (n+2)\alpha_n^m\} &= 2\int_{\delta-r_2}^{r_1}(r_1^2 - r^2)^{n/2} \times \\
&\quad \times \frac{d}{dr}\left\{r^mI_{-1,0}^m(\cos\theta_0)\right\}r^2\,dr. \tag{2.13}
\end{aligned}
$$

If, furthermore, we make use of the fact that

$$\frac{d}{dr}\left\{r^m I^m_{-1,0}\right\} = m r^{m-1} I^m_{-1,0} + r^m \frac{dI^m_{-1,0}}{dr} \qquad (2.14)$$

and insert this on the right-hand side of (2.13), this latter equation assumes the form

$$r_1^{m+n+2}\{(m+n+2)\alpha^m_n - n\alpha^m_{n-2} =$$
$$= -2 \int_{\delta-r_2}^{r_1} (r_1^2 - r^2)^{n/2} \frac{dI^m_{-1,0}}{dr} r^{m+2}\, dr . \qquad (2.15)$$

By virtue of the fact that

$$\frac{dI^m_{-1,0}}{dr} = \cos^m \theta_0 \frac{d\theta_0}{dr} , \qquad (2.16)$$

where the angle θ_0 continues to be defined by Eq. (2.2), and Equation (2.15) can be rewritten as

$$(m+n+2)\alpha^m_n - n\alpha^m_{n-2} = 2\Im^m_{-1,n} , \qquad (2.17)$$

valid for $m = 0, 1, 2, \ldots$ and $n = 0, 1, 2, \ldots$; where, for $n \to 0$, $\lim n\alpha^m_{n-2}$ $(m = 0, 1)$ is as given by Eqs. (3.46)–(3.47) of Chapter IV, and

$$\pi r_1^{m+n+2} \Im^m_{-1,n} = (r_2^2 - \delta^2) I^m_{0,-1,n} + \delta I^{m+1}_{0,-1,n} . \qquad (2.18)$$

Equations (2.11) and (2.17) represent two fundamental and mutually independent recursion formulae for associated α-functions, whose right-hand sides consist of integrals of the form $I^m_{0,1,n}$ or $\Im^m_{-1,n}$. Moreover, it can be seen that

$$2\delta I^m_{0,\beta,\gamma} = 2\delta s\, I^{m-1}_{0,\beta,\gamma} - I^{m-1}_{0,\beta,\gamma+2} \qquad (2.19)$$

and

$$2\delta r_1 \Im^m_{-1,n} = 2\delta s\, \Im^{m-1}_{-1,n} - r_1^2 \Im^{m-1}_{-1,n+2} . \qquad (2.20)$$

The first, combined with (2.11) leads to an 8-term recursion formula for the associated α-functions of the form

$$2\delta r_1\{(m+2)[\alpha^{m+1}_{n+2} - \alpha^{m+1}_n] + (m+3)\alpha^{m+3}_n\} -$$
$$- 2\delta s\{(m+1)[\alpha^m_{n+2} - \alpha^m_n] + (m+2)\alpha^{m+2}_n\} +$$
$$+ r_1^2\{(m+1)[\alpha^m_{n+4} - \alpha^m_{n+2}] + (m+2)\alpha^{m+2}_{n+2}\} = 0 ; \qquad (2.21)$$

while a combination of (2.18) with (2.20) furnishes a 5-term recursion formula

$$2\delta r_1\{m+n+5)\alpha^{m+1}_{n+2} - (n-2)\alpha^{m+1}_n\} -$$
$$- 2\delta s\{(m+n+4)\alpha^m_{n+2} - (n+2)\alpha^m_n\} +$$
$$+ r_1^2\{(m+n+6)\alpha^m_{n+4} - (m+4)\alpha^m_{n+2}\} = 0 ; \qquad (2.22)$$

both valid for $m \geq 0$ and $n \geq -1$, which consist only of associated α-functions of different orders and indices, with coefficients representing algebraic functions of the elements of the eclipse. In particular, Equation (2.11) with $m = 0$ discloses immediately that

$$r_1^{n+3} \alpha_n^1 = 2\delta r_2^{n+2} I_{1,n}^0 = \frac{2r_2^{n+3}}{n+2} I_{-1,n+2}^1 , \tag{2.23}$$

where the functions $I_{\beta,\gamma}^m$ continue to be given by Eq. (1.34).

Turning to the latter family of functions and setting

$$I_{\beta,\gamma}^m \equiv (\delta/r_2)^{\gamma/2} J_{\beta,\gamma}^m \tag{2.24}$$

we readily find the J-functions so defined to satisfy the recursion formulae

$$J_{\beta+2,\gamma}^m = J_{\beta,\gamma}^m - J_{\beta,\gamma}^{m+2} \tag{2.25}$$

and

$$J_{\beta,\gamma+2}^m = 2\left\{ J_{\beta,\gamma}^{m+1} - \mu J_{\beta,\gamma}^m \right\} , \tag{2.26}$$

where

$$\mu \equiv \frac{\delta - s}{r_2} = \frac{r_2^2 - r_1^2 + \delta^2}{2\delta r_1} . \tag{2.27}$$

The proofs of (2.25) and (2.26) are trivial (by vritue of the identity of their integrands); but that of

$$\gamma J_{\beta+2,\gamma-2}^m = (\beta + m + 2) J_{\beta,\gamma}^{m+1} - m J_{\beta,\gamma}^{m-1} \tag{2.28}$$

is not; and was discovered only recently by Lanzano (1976b).

The foregoing Equations (2.25)–(2.28) are the only ones known to relate three $J_{\beta,\gamma}^m$'s for different values of β, γ and m in an algebraic manner. Four-term recursion formulae are many, and have been exhaustively studied by the same investigator (cf. Lanzano, 1976a,b,c; for their summary cf. section III-3 of Kopal, 1979) with the aid of the recursion formulae satisfied by Appell's generalized hypergeometric series (cf. the next section of this chapter). Recursion formulae for the $J_{\beta,\gamma}^m$'s involving 5 or more terms are likewise known to exist; but being of lesser practical importance they will not be reviewed in this place.

In order to investigate such properties, let us depart from the fact that the associated alpha-functions α_n^m as well as the boundary integrals of the form $I_{\beta,\alpha}^m$ and $\mathfrak{S}_{\beta,\gamma}^m$, as defined in this chapter, depend on the parameters r_1, r_2 and δ only through their *ratios* and are, therefore *homogeneous* in them of zero degree. Accordingly, Euler's theorem on homogeneous functions leads us to anticipate that they will satisfy the partial differential equation

$$\left\{ r_1 \frac{\partial}{\partial r_1} + r_2 \frac{\partial}{\partial r_2} + \delta \frac{\partial}{\partial \delta} \right\} \left\{ \alpha_n^m, I_{\beta,\gamma}^m, \mathfrak{S}_{\beta,\gamma}^m \right\} = 0 \tag{2.29}$$

for any values of m, n or β and γ for which these functions are bounded.

This equation represents, to be sure, merely an algebraic relation obtaining between the first partial derivatives of the respective functions. In order to detail these in a more specific form, let us return to Eq. (2.1) for α_n^m and differentiate both sides of it with respect to r_1: in doing so we find that

$$r_1^{m+n+1}\left\{(m+n+2)\alpha_n^m + r_1\frac{\partial \alpha_n^m}{\partial r_1}\right\} =$$

$$= 2nr_1\int_{\delta-r_2}^{r_1}(r_1^2-r^2)^{\frac{n-2}{2}}I_{-1,0}^m(\cos\theta_0)r^{m+1}\,dr \equiv$$

$$\equiv nr_1^{m+n+1}\alpha_{n-2}^m,\qquad(2.30)$$

yielding

$$(m+n+2)\alpha_n^m - n\alpha_{n-2}^m \equiv -r_1\frac{\partial \alpha_n^m}{\partial r_1} = 2\Im_{-1,n}^m\qquad(2.31)$$

by (2.17).

Differentiating Eq. (2.5) next with respect to r_2, we find that

$$r_1^{m+n+2}\frac{\partial \alpha_n^m}{\partial r_2} = 2\int_{\delta-r_2}^{r_1}(r_1^2-r^2)^{n/2}\frac{\partial I_{-1,0}^m}{\partial r_2}r^{m+1}\,dr\qquad(2.32)$$

where, by Eqs. (2.3) and (2.4),

$$\pi\frac{\partial I_{-1,-0}^m}{\partial r_2} = \cos^m\theta_0\frac{\partial \theta_0}{\partial r_2} = \left\{\frac{\delta^2+r^2-r_2^2}{2\delta r}\right\}^m\frac{2r_2}{\sqrt{(2\delta r)^2-(\delta^2+r^2-r_2^2)^2}}.$$
$$(2.33)$$

Therefore, on introducing a variable x defined by

$$\delta^2+r^2-r_2^2 = 2\delta x,\qquad(2.34)$$

so that $r_1^2-r^2 = 2\delta(s-x)$ and $\delta^2+r_2^2-r^2 = 2\delta(\delta-x)$, we can rewrite (2.33) as

$$\pi r_1^{m+n+2}\frac{\partial \alpha_n^m}{\partial r_2} = 2r_2\int_{\delta-r_2}^{s}[r_2^2-(\delta-x)^2]^{-1/2}[2\delta(s-x)]^{n/2}x^m\,dx \equiv 2r_2\,I_{0,-1,n}^m$$
$$(2.35)$$

in acordance with Eq. (1.27). Lastly, by invoking the use of Euler's theorem (2.29) and inserting from (2.19), (2.31) and (2.35), in (2.29) for α_n^m we find that

$$\pi r_1^{m+n+2}\delta\frac{\partial \alpha_n^m}{\partial \delta} \equiv 2\{\pi r_1^{m+n+2}\Im_{-1,n}^m - r_2^2 I_{0,-1,n}^m\} =$$

$$= 2\delta\{I_{0,-1,n}^{m+1} - \delta\,I_{0,-1,n}^m\}.\qquad(2.36)$$

Therefore, for all values of $m \geq 0$ and $n \geq -1$ the partial derivatives of $\alpha_n^m(r_1,r_2,\delta)$ with respect to its arguments can be expresed in a closed form in terms of the $I_{0,-1,n}^m$-integrals as

$$\pi r_1^{m+n+3}\frac{\partial \alpha_n^m}{\partial r_1} = -2\{(r_2^2-\delta^2)I_{0,-1,n}^m + \delta I_{0,-1,n}^{m+1}\} \equiv$$

$$\equiv -2\pi r_1^{m+n+2}\,\Im_{-1,n}^m,\qquad(2.37)$$

$$\pi r_1^{m+n+2} \frac{\partial \alpha_n^m}{\partial r_2} = 2 r_2 I_{0,-1,n}^m \,, \tag{2.38}$$

$$\pi r_1^{m+n+2} \frac{\partial \alpha_n^m}{\partial \delta} = 2\{ I_{0,-1,n}^{m+1} - \delta I_{0,-1,n}^m \} \,. \tag{2.39}$$

An elimination of the I's between these equations discloses, moreover, the existence of a differential recursion formula

$$r_1 \frac{\partial \alpha_n^{m+1}}{\partial r_2} = \delta \frac{\partial \alpha_n^m}{\partial r_2} + r_2 \frac{\partial \alpha_n^m}{\partial \delta} \,. \tag{2.40}$$

If, in particular, the index $m = 0$, the foregoing equations (2.37)–(2.39) reduce further to

$$r_1 \frac{\partial \alpha_n^0}{\partial r_1} = \frac{2}{r_2} \left(\frac{r_2}{r_1} \right)^{n+2} \left\{ \delta I_{-1,n}^1 - r_2 I_{-1,n}^0 \right\} \equiv -2 \Im_{-1,n}^0, \tag{2.41}$$

$$r_2 \frac{\partial \alpha_n^0}{\partial r_2} = 2 \left(\frac{r_2}{r_1} \right)^{n+2} I_{-1,n}^0 \,, \tag{2.42}$$

$$\delta \frac{\partial \alpha_n^0}{\partial \delta} = -2 \frac{\delta}{r_2} \left(\frac{r_2}{r_1} \right)^{n+2} I_{-1,n}^1 \,; \tag{2.43}$$

where the functions $I_{-1,n}^0$ and $I_{-1,n}^1$ on the right-hand sides of (2.41)–(2.43) represent particular cases of integrals defined by Eq. (1.34) corresponding to $\beta = -1$, $\gamma = n$, and $m = 0, 1$.

Next, let us normalize their limits to the interval $(0, 1)$ by introducing a new variable u defined as

$$\delta - x = r_2 (1 - 2\kappa^2 u) \,, \tag{2.44}$$

where the modulus

$$\kappa^2 = \frac{1}{2} \left(1 - \frac{\delta - s}{r_2} \right) = \frac{r_1^2 - (\delta - r_2)^2}{4 \delta r_2} \tag{2.45}$$

is identical with that previously introduced by Eqs. (1.24) or (1.49). If so, then for $m = 0$ or 1 (but any value of n!) we recognize in the corresponding functions $I_{-1,n}^m$ integral representations of ordinary hypergeometric series of the type $_2F_1$ of the form

$$2\pi I_{-1,n}^m = \left(\frac{\delta}{r_2} \right)^{n/2} B \left(\frac{1}{2}, \frac{n+2}{2} \right) (2\kappa)^{n+1} \times$$
$$\times {}_2F_1 \left(\frac{1}{2} - m, \frac{1}{2} + m, \frac{n+3}{2}; \kappa^2 \right) \tag{2.46}$$

if the eclipse is partial (B standing for the beta-function of the respective arguments), and

$$I_{-1,n}^m = \left(\frac{n}{4\kappa^2} \right)^m \left(\frac{\delta}{r_2} \right)^{n/2} (2\kappa)^n \, {}_2F_1 \left(m - \frac{n}{2}, m + \frac{1}{2}, 2m+1; \frac{1}{\kappa^2} \right) \tag{2.47}$$

if it is annular.

If the eclipse happens to be an occultation (i.e., $r_1 < r_2$), then the value of the modulus κ^2 is constrained by the inequality $0 \leq \kappa^2 \leq \frac{1}{2}$; while for transit eclipses ($r_1 > r_2$) it can be anywhere within $0 < \kappa^2 < 1$ (see Figure II.2) during partial eclipses, while during annular phase of transit eclipses $\kappa^2 > 1$.

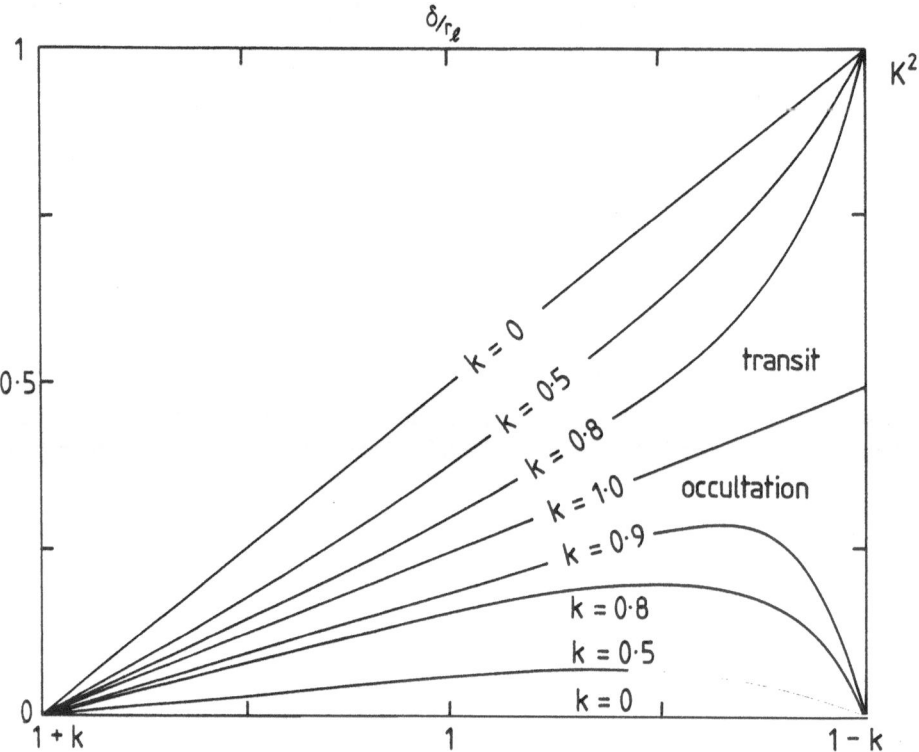

Figure II.2: Variation of the modulus κ^2 with δ during partial phases of the eclipse.

Accordingly, during *partial* phases of the eclipses,

$$\pi r_2 \frac{\partial \alpha_n^0}{\partial r_2} = \left(\frac{r_2}{r_1}\right)^{n+2} \left(\frac{\delta}{r_2}\right)^{n/2} B\left(\frac{1}{2}, \frac{n+2}{2}\right) (2\kappa)^{n+1} \times$$
$$\times {}_2F_1\left(\frac{1}{2}, \frac{1}{2}, \frac{n+3}{2}; \kappa^2\right) \tag{2.48}$$

and

$$\pi r_1 \frac{\partial \alpha_n^0}{\partial \delta} = -\left(\frac{r_2}{r_1}\right)^{n+1} \left(\frac{\delta}{r_2}\right)^{n/2} B\left(\frac{1}{2}, \frac{n+2}{2}\right) (2\kappa)^{n+1} \times$$

$$\times_2 F_1 \left(-\frac{1}{2}, \frac{3}{2}, \frac{n+3}{2}; \kappa^2 \right) ;$$ (2.49)

while if the eclipse becomes *annular*,

$$r_2 \frac{\partial \alpha_n^0}{\partial r_2} = 2 \left(\frac{r_2}{r_1} \right)^{n+2} \left(\frac{\delta}{r_2} \right)^{n/2} (2\kappa)_2^n F_1 \left(-\frac{n}{2}, \frac{1}{2}, 1; \frac{1}{\kappa^2} \right)$$ (2.50)

and

$$r_1 \frac{\partial \alpha_n^0}{\partial \delta} = -n \left(\frac{r_2}{r_1} \right)^{n+1} \left(\frac{\delta}{r_2} \right)^{n/2} (2\kappa)_2^{n-2} F_1 \left(\frac{2-n}{2}, \frac{3}{2}, 3; \frac{1}{\kappa^2} \right).$$ (2.51)

The reader may note that the hypergeometric series on the right-hand sides of Eqs. (2.48) and (2.49)—though *not* in (2.50)–(2.51)—represent spherical harmonics of fractional orders. In fact, Eqs. (2.48) and (2.49) can be rewritten alternatively as

$$\delta \frac{\partial \alpha_n^0}{\partial r_2} = \frac{\Gamma(1 + \frac{1}{2})n}{\sqrt{2\pi}} \left(\frac{2\delta r_2}{r_1^2} \right)^{\frac{n+2}{2}} (1 - \mu^2)^{\frac{n+1}{4}} P_{-\frac{1}{2}}^{-\frac{n+1}{2}} (\mu)$$ (2.52)

and

$$\delta \frac{\partial \alpha_n^0}{\partial \delta} = -\frac{\Gamma(1 + \frac{1}{2}n)}{\sqrt{2\pi}} \left(\frac{2\delta r_2}{r_1^2} \right)^{\frac{n+2}{2}} (1 - \mu^2)^{\frac{n+1}{4}} P_{\frac{1}{2}}^{-\frac{n+1}{2}} (\mu),$$ (2.53)

where the $P_{\pm 1/2}^{-(n+1)/2}(\mu)$'s satisfy the Legendre differential equation

$$\frac{d}{d\mu} \left\{ (1 - \mu^2) \frac{d}{d\mu} \right\} P_{\pm\frac{1}{2}}^{-\frac{n+1}{2}} + \frac{1}{4} \left\{ 1 \pm 2 - \frac{(n+1)^2}{1 - \mu^2} \right\} P_{\pm\frac{1}{2}}^{-\frac{n+1}{2}} = 0$$ (2.54)

and

$$\mu \equiv \frac{\delta - s}{r_2} = \frac{r_2^2 - r_1^2 + \delta^2}{2\delta r_2} = 1 - 2\kappa^2 ,$$ (2.55)

in agreement with Eqs. (1.24) and (2.27).

II.3 Differential Properties of Eclipse Functions

In order to investigate the differential properties of functions of the type $I_{\beta,\gamma}^m$ for $m > 1$, let us return to Eq. (2.24), and change over from x to u by (2.44) so as to reduce the limits of integration to (0,1): in doing so we find that

$$2\pi J_{\beta,\gamma}^m = (2\kappa)^{\beta+\gamma+2} \int_0^1 u^{\beta/2} (1 - \kappa^2 u)^{\beta/2} \times$$
$$\times (1 - u)^{\gamma/2} (1 - 2\kappa^2 u)^m \, du ,$$ (3.1)

where the modulus κ^2 (through which $J_{\beta,\gamma}^m$ depends on the geometry of the eclipses) continues to be given by Equation (2.45). The aim of the substitution (2.24) has, in fact, been only to factor out the functions $I_{\beta,\gamma}^m$ of two variables into a product of two functions—each depending on a single variable (δ/r_2 or κ).

Moreover, the integral on the right-hand side of the foregoing equation (3.1) can be readily recognized as an integral representation of Appell's generalized hypergeometric series $F^{(1)}$, defined by

$$F^{(1)}(a; b, b'; c; x, y) = \sum_{m=0}^{\infty} \sum_{n=0}^{\infty} \frac{(a)_{m+n}(b)_m(b')_n}{(c)_{m+n}} \frac{x^m y^n}{m!\, n!}. \tag{3.2}$$

In fact, if we abbreviate

$$a = \frac{1}{2}(\beta + 2) \quad \text{and} \quad b = \frac{1}{2}(\beta + \gamma + 4), \tag{3.3}$$

Eq. (3.33) can be rewritten as

$$2\pi J_{\beta,\gamma}^m = (2\kappa)^{2(b-1)} B(a, b-a) F^{(1)}(a; 1-a, -m; b; \kappa^2, 2\kappa^2), \tag{3.4}$$

where $F^{(1)}$ denotes the first one of Appell's generalized hypergeometric functions (cf. Appell and Kampé de Fériet, 1926) in two variables $x = \kappa^2$ and $y \equiv 2\kappa^2$.

The function $F^{(1)}$ is known to satisfy, in general a certain system of two simultaneous partial differential equations of second order. If, however, as in the present case, the ratio x/y is constant, this system can be reduced to a single ordinary differential equation, which was obtained by Burchnall (1942). From his results we deduce that if t denotes the operator

$$t \equiv \kappa^2 \frac{d}{d\kappa^2}, \tag{3.5}$$

the function $F^{(1)}$ on the r.h.s. of Eq. (3.2) satisfies an ordinary differential equation

$$t(t + b - 1)(t + b - 2)F^{(1)} - \kappa^2(3t + 1 - a - 2m)(t + b - 1)(t + a)F^{(1)}$$

$$+ 2\kappa^4(t + 1 - a - m)(t + a)(t + a + 1)F^{(1)} = 0 \tag{3.6}$$

of third order. Moreover, a substitution

$$J_{\beta,\gamma}^m = \kappa^{2(b-1)} F^{(1)} \tag{3.7}$$

in (3.6) leads to

$$t(t - 1)(t + 1 - b)J_{\beta,\gamma}^m - \kappa^2(3t + 4 - a - 3b - 2m)t(t + a + 1 - b)J_{\beta,\gamma}^m$$

$$+ 2\kappa^4(t + 2 - a - b - m)(t + a + 1 - b)(t + a + 2 - b)J_{\beta,\gamma}^m = 0, \tag{3.8}$$

disclosing that the function $J_{\beta,\gamma}^m$—and, therefore, the derivative $\partial \alpha_n^m / \partial \delta$ as given by Eqs. (2.39) and (1.33)–(1.34) with (2.24)—satisfies a linear differential equation of *third* order; and, if so, the function α_n^0 itself will satisfy such an equation of *fourth* order.

A solution of Eq. (3.8) for $J_{\beta,\gamma}^m$ can be sought as a series in ascending powers of κ^2. Since the operator t occurs as a factor of the first as well as the second term on the l.h.s. of (3.8), the unit difference between the roots of the respective indicial equation will not give rise to any logarithmic singularity at the origin. If we construct such a particular solution by standard methods, we find it to assume the form

$$2\pi J_{\beta,\gamma}^m = 4^{b-1} B(a, b-a) \sum_{j=0}^{\infty} \frac{a_j(1-a)_j}{j!(b)_j} {}_2F_1(-j, -m; a-j; 2)\kappa^{2(b+j-1)}, \quad (3.9)$$

where $(a)_j \equiv a(a+1)(a+2)\ldots(a+j-1)$, $(a)_0 = 1$ denotes the customary Pochhammer symbols, and $_2F_1(-j, -m, a-j.2)$ stands for a Jacobi polynomial of degree j. The foregoing series (3.9) constitutes the most general representation of the integral $J_{\beta,\gamma}^m$ valid for all (not necessarily integral) values of β, γ or m—provided only that their combination is such as to make the series on the r.h.s. of (3.9) convergent. Since, in our problem m is restricted to assume the values of zero or a positive integer, while $\beta + \gamma > -2$, the convergence of the expansion on the r.h.s. of Eq. (3.9) is assured.

It may also be noted that, for three particular values of the parameters β, γ and m, the differential equation (3.6) governing the function $F^{(1)}$ reduces to one of second order. This will happen first if $m = 0$; for the expansion on the r.h.s. of (3.9) reduces then to an ordinary hypergeometric series; and, as a result,

$$2\pi J_{\beta,\gamma}^m = 4^{b-1} B(a, b-a)\kappa^{2(b-1)} F(a, 1-a, b, \kappa^2), \quad (3.10)$$

in conformity with Eq. (2.46). The second reducible case arises when $\gamma = 0$ and, therefore, $a = b - 1$. If so, Eq. (3.6) can be reduced (by removal of a factor) to

$$\{t - \kappa^2(3t + 2 - b - 2m) + 2\kappa^4(t + 2 - b - m)\}(t + b - 1)F^{(1)} = 0, \quad (3.11)$$

which is also one of second order (cf. Heun, 1889). A third reducible case arising when $2(\beta + m + 2) + \gamma = 0$ (Chaundy, 1943) has no relevance to our applications, and is being mentioned only for the sake of completeness.

All equations given so far hold good only for partial eclipses. Should the latter become *annular*, the substitution required to reduce $J_{\beta,\gamma}^m$ to tractable form becomes

$$\delta - x = r_2(1 - 2v) \quad (3.12)$$

in place of (2.44) used formerly; and

$$\pi J_{\beta,\gamma}^m = 2^{\beta+\gamma+1}\kappa^\gamma \int_0^1 v^{\beta/2}(1-v)^{\beta/2}(1-\kappa^{-2}v)^{\gamma/2}(1-2v)^m dv; \quad (3.13)$$

and the integral on the r.h.s. represents once more Appell's generalized hypergeometric series of the form

$$2\pi J_{\beta,\gamma}^m = 4^{b-1} B(a, a)\kappa^{2(b-a-1)} F^{(1)}(a; -m, a-b+1; 2a; 2, \kappa^{-2}). \quad (3.14)$$

The differential equation governing this function is again of the form (3.8), and can be obtained from it by an appropriate permutation of parameters. Its solution is, in turn, expressible as

$$2\pi J_{\beta,\gamma}^m = 4^{b-1} B(a,a) \sum_{j=0}^{\infty} \frac{(a)_j (a-b+1)_j}{j!(2a)_j} \, _2F_1(-m, a+j, 2a+j, 2)\kappa^{2(b-a-j-1)}.$$

(3.15)

If $m = 0$, this series reduces to

$$2\pi J_{\beta,\gamma}^m = 4^{b-1} B(a,a)\kappa_2^{2(b-a-1)} F_1(a, a-b+1, 2a; \kappa^{-2});$$

(3.16)

while if, in addition, $\gamma = 0$ (i.e., $a = b - 1$),

$$2\pi J_{\beta,\gamma}^m = 4^a B(a,a).$$

(3.17)

This latter expression no longer involves κ^2 and, therefore, does not vary during annular phase.

For $m = 0$ the differential equation (2.43) together with (2.46) or (2.47) can be considerably simplified. For differentiating the latter with respect to δ we find that

$$\pi r_1 \frac{\partial^2 \alpha_n^0}{\partial \delta^2} = -\left(\frac{r_2}{r_1}\right)^{n+1} B\left(\frac{1}{2}, \frac{n+2}{2}\right) \frac{\partial}{\partial \delta}\left\{\left(\frac{\delta}{r_2}\right)^{n/2} (2\kappa)^{n+1} \times \right.$$

$$\left. \times {}_2F_1\left(-\frac{1}{2}, \frac{3}{2}, \frac{n+3}{2}; \kappa^2\right)\right\}.$$

(3.18)

Since, moreover, by a well-known hypergeometric identity,

$$\frac{\partial}{\partial \kappa^2}\left\{\kappa_2^{n-1} F_1\left(-\frac{1}{2}, \frac{3}{2}, \frac{n+3}{2}; \kappa^2\right)\right\} =$$

$$= \frac{n+1}{2}\kappa^{n-1} \, _2F_1\left(-\frac{1}{2}, \frac{3}{2}, \frac{n+1}{2}; \kappa^2\right),$$

(3.19)

it can be shown that, for $n > 0$, the functions satisfy a linear differential recursion formula of the form

$$\frac{\partial^2 \alpha_n^0}{\partial \delta^2} - \frac{n}{2\delta}\frac{\partial \alpha_n^0}{\partial \delta} + \frac{n}{2}\frac{\delta^2 + r_1^2 - r_2^2}{\delta r_1^2}\frac{\partial \alpha_{n-2}^0}{\partial \delta} = 0$$

(3.20)

of second order, valid for *any* type of eclipse (be it partial or annular); while a similar differentiation of (2.41) and (2.42) with respect to r_1 and r_2 discloses that

$$\frac{\partial^2 \alpha_n^0}{\partial r_1^2} + \frac{n+3}{r_1}\frac{\partial \alpha_n^0}{\partial r_1} - \frac{n}{r_1}\frac{\partial \alpha_{n-2}^0}{\partial r_1} = 0$$

(3.21)

and

$$\frac{\partial^2 \alpha_n^0}{\partial r_2^2} - \frac{n+2}{2r_2}\frac{\partial \alpha_n^0}{\partial r_2} - \frac{n}{2r_2}\frac{\delta^2 - r_1^2 - r_2^2}{r_1^2}\frac{\partial \alpha_{n-2}^0}{\partial r_2} = 0.$$

(3.22)

Since, however, by Euler's theorem for homogeneous functions, the partial first derivatives of α_n^0 with respect to r_1, r_2 and δ are related by Equation (2.29), only two of the three equations (3.20)–(3.22) are mutually independent.

The latter equations represent linear differential relations between two associate α-functions of orders n and $n-2$; but α_{n-2}^0 can be eliminated from them by the following strategy. First, let us resort to the recursion formula (2.31), which on insertion from (2.29) discloses that

$$n\alpha_{n-2}^0 = (n+2)\alpha_n^0 + r_1\frac{\partial\alpha_n^0}{\partial r_1} =$$

$$= (n+2)\alpha_n^0 - r_2\frac{\partial\alpha_n^0}{\partial r_2} - \delta\frac{\partial\alpha_n^0}{\partial\delta}. \tag{3.23}$$

An elimination of the terms $n\alpha_{n-2}^0$ from Equations (3.21)–(3.22) by means of (3.23) permits us to rewrite the former as

$$\frac{\partial}{\partial r_1}\left(r_1\frac{\partial\alpha_n^0}{\partial r_1}\right) + r_2\frac{\partial^2\alpha_n^0}{\partial r_2\partial r_1} + \delta\frac{\partial^2\alpha_n^0}{\partial\delta\partial r_1} = 0, \tag{3.24}$$

$$\frac{\partial^2\alpha_n^0}{\partial r_2^2} - \frac{n+2}{2r_2}\left(\frac{\delta^2 - r_2^2}{r_1^2}\right)\frac{\partial\alpha_n^0}{\partial r_2} - \frac{\delta^2 - r_1^2 - r_2^2}{2r_1r_2}\frac{\partial^2\alpha_n^0}{\partial r_1\partial r_2} = 0, \tag{3.25}$$

$$\frac{\partial^2\alpha_n^0}{\partial\delta^2} + \frac{1}{\delta}\left(1 + \frac{n+2}{2}\frac{\delta^2 - r_2^2}{r_1^2}\right)\frac{\partial\alpha_n^0}{\partial\delta} + \frac{\delta^2 + r_1^2 - r_2^2}{2\delta r_1}\frac{\partial^2\alpha_n^0}{\partial\delta\partial r_1} = 0, \tag{3.26}$$

which differ from (3.20)–(3.22) insofar as they contain but one and the same dependent, but two or three independent variables.

Let us conclude this section by reducing the simultaneous equations governing α_n^0 to a linear system of four partial differential equations of first order, which is more suitable than others for the purposes of numerical integration. In order to do so, let us recall that the associated alpha-functions α_n^0 is a homogeneous function, of zero order, of three independent variables r_1, r_2, δ occurring in α_n^0 only through their squares; and can, therefore, be made to depend only on the ratios (say)

$$(r_2/r_1)^2 \equiv x \text{ and } (\delta/r_1)^2 \equiv y. \tag{3.27}$$

Let, moreover,

$$\frac{\partial\alpha_n^0}{\partial x} \equiv p, \quad \frac{\partial\alpha_n^0}{\partial y} \equiv q \tag{3.28}$$

and

$$\frac{\partial^2\alpha_n^0}{\partial x^2} \equiv r, \quad \frac{\partial^2\alpha_n^0}{\partial x\partial y} \equiv s, \quad \frac{\partial^2\alpha_n^0}{\partial y^2} \equiv t \tag{3.29}$$

stand for partial derivatives of α_n^0 with respect to x and y. If so, Equations (3.20)–(3.22) can be rewritten in terms of the notations (3.27)–(3.29) as

$$(1-y)yt - x^2r - 2xys - (1-\nu)(xp + yq) + q = 0, \tag{3.30}$$

$$(1 - x + y)xr - (1 + x - y)ys - (1 - \nu)(x - y)p = 0, \qquad (3.31)$$

$$(1 + x - y)yt - (1 - x + y)xs + (1 - \nu)(x - y)q + q = 0; \qquad (3.32)$$

where we have abbreviated

$$\nu = \frac{n + 2}{2}. \qquad (3.33)$$

In order to solve these equations, let us subtract (3.30) from (3.32) and divide by x: the result discloses that

$$xr + yt - (1 - x - y)s + (1 - \nu)(p + q) = 0. \qquad (3.34)$$

On the other hand, a subtraction of (3.32) from (3.31) yields

$$xr - yt - q - (x - y)\{xr + yt - (1 - x - y)s + (1 - \nu)(p + q)\} = 0, \qquad (3.35)$$

which on insertion from (3.34) reduces to

$$xr - yt = q \qquad (3.36)$$

an equation which is independent of n (i.e., ν).

Therefore, the simplest simultaneous system of *two* independent equations representing (3.30)–(3.32) can be written as

$$\left. \begin{array}{rl} xr + yt + (1 - \nu)(p + q) &= (1 - x - y)s, \\ xr - yt &= q, \end{array} \right\} \qquad (3.37)$$

which in the case of uniformly bright discs ($n = 0$ and, therefore, $\nu = 1$) reduces to

$$\left. \begin{array}{rl} xr + yt &= (1 - x - y)s, \\ xr - yt &= q. \end{array} \right\} \qquad (3.38)$$

Another—more symmetrical—combination of (3.30)–(3.32) obtains if Eq. (3.37) multiplied by y is subtracted from (3.31): the result adjoined to (3.30) leads to the simultaneous system

$$\left. \begin{array}{rl} (1 - x)xr - y^2 t - 2xys - (1 - \nu)(xp + yq) &= 0, \\ (1 - y)yt - x^2 r - 2xys - (1 - \nu)(xp + yq) + q &= 0, \end{array} \right\} \qquad (3.39)$$

of equations which are known to be satisfied by Appell's generalized hypergeometric series $F^{(4)}(\alpha, \beta; \gamma, \gamma'; x, y)$ of the fourth kind, in two variables. In fact, we established already (cf. Kopal, 1977b; or Eq. (3.77) on p.38 of Kopal, 1979) that, for *annular* eclipses (in our present notations)

$$\alpha_n^0(x, y) = xF^{(4)}(1 - \nu, 1; 2, 1; x, y), \qquad (3.40)$$

which satisfies the system (3.39) and can, therefore, be regarded as one of its particular solutions.

The same is true of the solution

$$\alpha_n^0(x, y) = \frac{2}{n+2} \equiv \frac{1}{\nu} = \text{constant}, \tag{3.41}$$

obtaining in the course of *total* eclipses (when all terms on the left-hand sides of Eqs. (3.39) vanish identically); but to construct the expressions appropriate for *partial* eclipses calls for a combination of more than one particular solution of (3.39). The system itself being one of fourth order, its complete primitive will consist of a linear combination of four particular solutions of (3.39); and since, in our case, the coefficients β as well as γ and γ' in the series for $F^{(4)}$ are integers, all three remaining particular solutions of (3.39) possess logarithmic singularities. For fuller details of their actual forms the reader is referred, e.g., to p.52 of the treatise by Appell and Kampé de Fériet (1926); but these appear to be too complicated to be of much practical use.

Next, we wish to point out the way in which the system of partial differential equation (3.37) can be used to construct its solution for constant (or otherwise prescribed) values of x. The total differential of the function $\alpha_n^0(x, y)$ can be written as

$$d\alpha_n^0 = \frac{\partial \alpha_n^0}{\partial x}dx + \frac{\partial \alpha_n^0}{\partial y}dy \equiv p\,dx + q\,dy \,; \tag{3.42}$$

and similar subsequent differentiation yields

$$dp = \frac{\partial p}{\partial x}dx + \frac{\partial p}{\partial y}dy \equiv r\,dx + s\,dy, \tag{3.43}$$

$$dq = \frac{\partial q}{\partial x}dx + \frac{\partial q}{\partial y}dy \equiv s\,dx + t\,dy, \tag{3.44}$$

and

$$ds = \frac{\partial s}{\partial x}dx + \frac{\partial s}{\partial y}dy \,. \tag{3.45}$$

Equations (3.37) can be solved for r and t as

$$r = a_1 s + a_2 p + a_3 q \,, \tag{3.46}$$

$$t = b_1 s + b_2 p + b_3 q \,, \tag{3.47}$$

where

$$a_1 = \frac{1 - x - y}{2x}, \quad a_2 = \frac{\nu - 1}{2x}, \quad a_3 = \frac{\nu}{2x} \tag{3.48}$$

and

$$b_1 = \frac{1 - x - y}{2y}, \quad b_2 = \frac{\nu - 1}{2y}, \quad b_3 = \frac{\nu - 2}{2y}. \tag{3.49}$$

The solutions (2.101)–(2.102) of (2.94) will be unique provided that

$$a_1 b_1 \equiv \frac{(1 - x - y)^2}{4xy} \neq 1 \,; \tag{3.50}$$

which is true at any phase, except when $\delta = |r_1 \pm r_2|$—i.e., at the moment of the first contact of the eclipse, and at second contact (internal tangency) marking the commencement of totality (for $r_2 > r_1$) or annular phase (if $r_2 < r_1$).

In order to determine the partial derivatives of s with respect to x and y occurring on the right-hand side of Eq. (3.45), differentiate Eq. (3.46) with respect to y, and Eq. (3.47) with respect to x: in doing so and noting that, by Eqs. (3.29),

$$\frac{\partial r}{\partial y} \equiv \frac{\partial s}{\partial x} \quad \text{and} \quad \frac{\partial t}{\partial x} \equiv \frac{\partial s}{\partial y} \,, \tag{3.51}$$

we find that

$$\frac{\partial s}{\partial x} = \frac{\partial a_1}{\partial y}s + a_1\frac{\partial s}{\partial y} + \frac{\partial u_2}{\partial y}p + a_2 s + \frac{\partial u_3}{\partial y}q + a_3 t \tag{3.52}$$

and

$$\frac{\partial s}{\partial y} = \frac{\partial b_1}{\partial x}s + b_1\frac{\partial s}{\partial x} + \frac{\partial b_2}{\partial x}p + b_2 r + \frac{\partial b_3}{\partial x}q + b_3 s, \tag{3.53}$$

where the coefficients a_j and b_j on the right-hand sides of Eqs. (3.46)–(3.47) continue to be given by (3.48)–(3.49).

On insertion for these in (3.52)–(3.53) we note thsat the latter equations constitute a simultaneous algebraic system for the determination of the desired partial derivatives of s. On evaluation these are found to be expressible as

$$\frac{\partial s}{\partial x} = \alpha_1 s + \alpha_2 p + \alpha_3 q \,, \tag{3.54}$$

$$\frac{\partial s}{\partial y} = \beta_1 s + \beta_2 p + \beta_3 q \,, \tag{3.55}$$

with the constants α_j and β_j $(j = 1, 2, 3)$ given by the following equations:

$$2Dx\alpha_1 = (\nu - 1)\{(1 - x - y)^2 + 4x(1 - x)\} - 2x(1 - x + y), \tag{3.56}$$

$$2Dx\alpha_2 = (\nu - 1)\{(\nu - 1)(1 + x - y) + 2x\} \,, \tag{3.57}$$

$$2Dx\alpha_3 = \nu\{(\nu - 1)(1 + x - y) - 2x\} \,; \tag{3.58}$$

and

$$2Dy\beta_1 = \nu\{(1 - x - y)^2 + 4x(1 - x)\} - 6y(1 + x - y), \tag{3.59}$$

$$2Dy\beta_2 = (\nu - 1)\{(\nu - 1)(1 - x + y) + (1 - x - y)\} \,, \tag{3.60}$$

$$2Dy\beta_3 = \nu\{(\nu - 1)(1 - x + y) - (1 - x - y)\} \,, \tag{3.61}$$

where

$$D \equiv 4xy - (1 - x - y)^2 \tag{3.62}$$

stands for a determinant which does not vanish if $\delta \neq |r_1 \pm r_2|$.

Suppose now that we wish to evaluate the function $\alpha_n^0(x, y)$ for *constant* value of x. If so, then obviously we are entitled to set $dx = 0$ on the right-hand sides of Eqs. (3.42)–(3.45), which reduces the latter system to four first-order *ordinary* differential equations of the form

$$\frac{d\alpha_n^0}{dy} = q, \tag{3.63}$$

$$\frac{dp}{dy} = s, \tag{3.64}$$

$$\frac{dq}{dy} = b_1 s + b_2 p + b_3 q, \tag{3.65}$$

$$\frac{ds}{dy} = \beta_1 s + \beta_2 p + \beta_3 q. \tag{3.66}$$

An elimination of p and s between Eqs. (3.64)–(3.66) leads to a *third*-order equation for $q \equiv \partial \alpha_n^0 / \partial y$; and once the latter has been integrated, α_n^0 follows from (3.63) by a mere quadrature of q. Such a third-order equation has already been obtained earlier in this chapter admitting of solution in terms of hypergeometric series; and Equations (2.41)–(2.47) disclosed that

$$2\nu B \left(\frac{1}{2}, \nu + \frac{1}{2}\right) q = (xy)^{\nu/2} (4\kappa^2)^{\nu - \frac{1}{2}} {}_2F_1 \left(-\frac{1}{2}, \frac{3}{2}; \nu + \frac{1}{2}; \kappa^2\right) \tag{3.67}$$

if the eclipse is partial, and

$$yq = (1 - \nu)(xy)^{\nu/2} (4\kappa^2)^{\nu - 2} {}_2F_1\left(2 - \nu, \frac{3}{2}; 3; \kappa^{-2}\right), \tag{3.68}$$

if it is annular; where the modulus

$$\kappa^2 = \frac{r_1^2 - (\delta - r_2)^2}{4\delta r_2} = \frac{1}{2} + \frac{1 - x - y}{4\sqrt{xy}}. \tag{3.69}$$

For κ^2 close to unity (i.e., in the neighbourhood of internal tangency) the convergence of the hypergeometric series on the right-hand sides of the preceding equations becomes, however, rather slow; and the equivalent set of first-order equations (3.64)–(3.66) is simpler to program and integrate on automatic computers, starting from the initial conditions

$$p_0 = (1 - x)^{\nu - 1}, \tag{3.70}$$

$$q_0 = (1 - \nu)x(1 - x)^{\nu - 1}, \tag{3.71}$$

$$s_0 = (1 - \nu)(1 - x)^{\nu - 3}\{1 + (1 - \nu)x\}, \tag{3.72}$$

following from Eq. (2.24) of Chapter III for $y = 0$ (cf. also Equation (2.26) of that chapter) when

$$\alpha_0 = \frac{1 - (1 - \nu)^\nu}{\nu} \; ; \tag{3.73}$$

and results of the integration can be checked by the requirement that, at the moment of internal tangency of a transit eclipse—when $\delta = r_1 - r_2$ and, therefore, $y_1 = (1 - \sqrt{x})^2$,

$$\alpha_n^0(y_1) = \frac{(2\nu)! \, x^{(\nu+1)/2}}{\Gamma(\nu+1)\Gamma(\nu+2)} \, {}_2F_1(1 - \nu, \nu + 1; \nu + 2; \sqrt{x}) =$$

$$= x^{(\nu+1)/2} \frac{{}_2F_1(1 - \nu, \nu + 1; \nu + 2; \sqrt{x})}{{}_2F_1(\nu - 1, \nu; 2\nu + 1; 1)} . \tag{3.74}$$

It may be added that all $\alpha_n^0(y_1)$'s as defined by the above equation are expressible in a closed form in terms of radicals and inverse trigonometric functions; for their explicit forms for $n = 0(1)4$ see Appendix I of the author's 1979 book.

II.4   Bibliographical Notes

An analytical evaluation of the fractional loss of light $\frac{3}{2}\alpha_1^0$ of discs linearly darkened at the limb in terms of elliptic integrals has first been given by Tsesevich (1936), and tabulated by him for different types of eclipses three years later (Tsesevich, 1939). Subsequently, the associated alpha-functions α_n^0 ($m = 0(1)4$ and $n = -1(1)4$) was explicitly evaluated by Kopal (1942b).

For a general discussion of eclipsing functions of the type α_n^m and I_β, γ^m and of their different properties cf. Kopal (1947) while the effects of limb-darkening of arbitrary degree were first investigated by Kopal (1949), a summary of which can be found in Chapter IV of the author's treatise on *Close Binary Systems* (Kopal, 1959; augmented twenty years later in his *Language of the Stars* (Kopal, 1979). For a more recent discussion of differential properties of such functions cf. also Kopal (1982d).

Chapter III

LOSS OF LIGHT AS INTEGRAL TRANSFORMS

In the preceding chapter of this book we have expressed the fractional loss of light due to eclipses of spherical stars arbitrarily darkened at the limb in terms of elementary geometry of the areas eclipsed, weighted in proportion of their relative brightness. However, this definition can be generalized, and expressed in more symmetrical form, by re-formulating our problem in a basically different way: namely, by regarding the fractional loss of light as a cross-correlation of two circular apertures—one representing the star undergoing eclipse, and the other as the eclipsing disc. In doing so, we shall be able to relate the loss of light arising from mutual stellar eclipses with the 'diffraction patterns' of apertures representing the two components—with arbitrary distribution of light over the star undergoing eclipse, or arbitrary transparency of the eclipsing object—and by explicit use of diffraction integrals in our analysis we shall be in a position to connect our subject more closely with the relevant parts of physical optics.

III.1 Beam Cross-Correlation: Fourier Transforms

In order to approach our subject in this manner, let us revert (temporarily) to the cosmology of the ancient Greeks and consider the stars to represent 'apertures' in the firmament of the Heavens, through which we can see the 'eternal fire'; of brightness yet to be specified. Let x, y be the rectangular coordinates in the plane of the plane tangent to the celestial sphere, with the origin at the centre of our aperture; and let, moreover, the function $f(x, y)$ represent the distribution of brightness within this aperture. If so, the two-dimensional Fourier transform $F(u, v)$ of $f(x, y)$ is known to be given by

$$F(u, v) = \iint_{-\infty}^{+\infty} f(x, y)e^{-2\pi i(xu+yv)} \, dx \, dy; \tag{1.1}$$

and if, moreover, the distribution of brightness $f(x, y)$ within the aperture is radially-symmetrical—so that

$$f(x, y) \equiv f(r) \tag{1.2}$$

where

$$x + iy = r\,e^{i\theta}, \tag{1.3}$$

41

it follows from (1.1) that

$$F(q, \phi) = \int_0^\infty \int_{-\pi}^{\pi} f(r) e^{-2\pi i qr \cos(\theta - \phi)} r \, dr \, d\theta, \qquad (1.4)$$

where

$$u + iv = q e^{i\phi}. \qquad (1.5)$$

In order to evaluate the integrals on the right-hand side of Equation (1.4), let us invoke the use of Jacobi's well-known expansion theorem (cf., e.g., Watson, 1945; p.22) which permits us to assert that

$$e^{-2\pi i qr \cos(\theta - \phi)} = e^{-2\pi i qr \sin(\frac{1}{2}\pi + \theta - \phi)} =$$

$$= J_0(2\pi qr) + 2 \sum_{n=1}^{\infty} J_{2n}(2\pi qr) \cos\{n\pi + 2n\theta - 2n\phi\} -$$

$$- 2i \sum_{n=1}^{\infty} J_{2n+1}(2\pi qr) \sin\{n\pi + \frac{1}{2}\pi +$$

$$+ (2n+1)(\theta - \phi)\}, \qquad (1.6)$$

where the symbols $J_n(x)$ denote the respective Bessel functions of the first kind, with real arguments, and $i \equiv \sqrt{-1}$ stands (as before) for the imaginary unit.

Since, however, for $n > 0$

$$\int_{-\pi}^{\pi} \cos 2n(\frac{1}{2}\pi + \theta - \phi) d\theta = 0, \qquad (1.7)$$

while

$$\int_{-\pi}^{\pi} \sin(2n + 1)(\frac{1}{2}\pi + \theta - \phi) d\theta = 0 \qquad (1.8)$$

for any value of n including zero, it follows that

$$\int_{-\pi}^{\pi} e^{-2\pi i qr \cos(\theta - \phi)} d\theta = 2\pi J_0(2\pi qr); \qquad (1.9)$$

and, accordingly,

$$F(q) = 2\pi \int_0^\infty f(r) J_0(2\pi qr) r \, dr, \qquad (1.10)$$

where, by (1.5), $q^2 = u^2 + v^2$. If, moreover, the 'aperture function' $f(r)$ vanishes for $r > r_1$, the foregoing equation can be rewritten as

$$F(q) = 2\pi \int_0^{r_1} f(r) J_0(2\pi qr) r \, dr; \qquad (1.11)$$

i.e., as a Hankel transform of $f(r)$ of zero order, which represents the well-known Airy diffraction pattern of a finite circular aperture of radius r_1 for a distribution of brightness $f(r)$ within that aperture (and zero outside of it).

Let us assume now that (in accordance with Equation (1.2) of Chapter II) this aperture function is given by a law of limb-darkening of the form

$$f(r) = f(0)\{1 - u_1 - u_2 - u_3 - \cdots + \sum_{n=1}^{N} u_n \cos^n \gamma\}, \qquad (1.12)$$

where u_1, u_2, u_3,... denote the respective coefficients of limb-darkening of the form

$$\cos \gamma = \frac{\sqrt{r_1^2 - r^2}}{r_1} \qquad (1.13)$$

represents the cosine of the angle of foreshortening (such that $r = r_1 \sin \gamma$). Since for any value of $\nu > 0$ (integral or fractional)

$$\int_0^{r_1} (r_1^2 - r^2)^{\nu-1} J_0(2\pi qr) r \, dr = 2^{\nu-1} \Gamma(\nu) \frac{J_\nu(2\pi qr)_1}{(2\pi qr_1)^\nu} r_1^{2\nu}, \qquad (1.14)$$

the diffraction pattern (1.11) of our limb-darkened circular 'aperture' can be expressed as

$$F(q) = L_1 \sum_{n=0}^{\infty} C^{(n)} 2^\nu \Gamma(\nu) \frac{J_\nu(2\pi qr_1)}{(2\pi qr_1)^\nu}, \qquad (1.15)$$

where the luminosity L_1 of the aperture and the coefficients $C^{(n)}$ dependent on its limb-darkening have been defined by Equations (1.6) and (1.11)–(1.12) of Chapter II, in which we have abbreviated

$$\nu = \frac{n+2}{2}; \qquad (1.16)$$

in accordance with (2.88 of Chapter II.

Let us now (cf. Figure III.1) turn our attention to the off-centre aperture which represents the eclipsing component, situated on the x-axis at a distance δ from the origin of coordinates. If so, the Fourier transform of this latter aperture should—by analogy with (1.1)—be defined by

$$G(u, v) = \int \int_{-\infty}^{+\infty} g(\xi, \eta) e^{-2\pi i[(\delta+\xi)u+\eta v]} d\xi \, d\eta, \qquad (1.17)$$

where

$$x \equiv \xi + \delta \quad \text{and} \quad y \equiv \eta. \qquad (1.18)$$

Let us, moreover, introduce again a set of plane polar coordinates ϱ, ζ related with the rectangular coordinates ξ, η by

$$\xi + i\eta = \varrho \, e^{i\zeta}; \qquad (1.19)$$

so that

$$dx \, dy = d\xi \, d\eta = \varrho \, d\varrho \, d\zeta. \qquad (1.20)$$

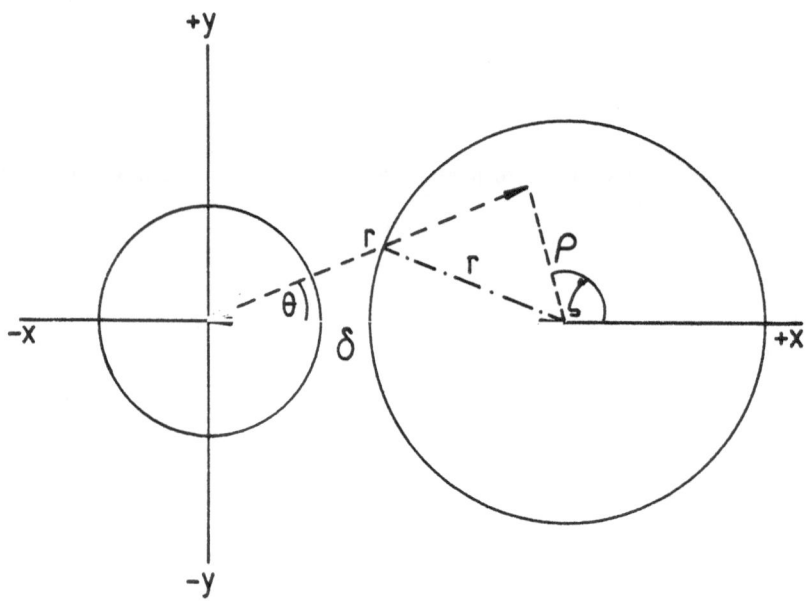

Figure III.1:

If so, then obviously

$$G(u, v) \; = \; e^{-2\pi i \delta u} \int_0^\infty \int_{-\pi}^{\pi} g(\varrho, \zeta) e^{-2\pi i q \varrho \, \cos(\zeta - \phi)} \, \varrho \, d\varrho \, d\zeta , \qquad (1.21)$$

where $g(\varrho, \zeta)$ denotes the 'transparency function' of the second aperture. If, in particular, the latter represents a circular disc which is wholly opaque for $\varrho \leq r_2$ and wholly transparent for $\varrho > r_2$, it would follow that

$$g(\varrho, \zeta) = \begin{cases} 1 & \text{for} \quad \varrho \leq r_2, \\ 0 & \text{for} \quad \varrho > r_2; \end{cases} \qquad (1.22)$$

in which case

$$G(u, v) \; = \; e^{-2\pi i \delta u} \int_0^{r_2} \varrho \, d\varrho \int_{-\pi}^{\pi} e^{2\pi i q \varrho \, \cos(\zeta - \phi)} \, d\zeta . \qquad (1.23)$$

Both integrals on the right-hand side of the foregoing equation can be evaluated in a closed form: for a resort to Equations (1.6)–(1.8) discloses readily that, since

$$2\pi q \varrho J_0(2\pi q \varrho) \; = \; \frac{d}{d\varrho} \{ \varrho J_1(2\pi q \varrho) \} \qquad (1.24)$$

and $J_1(0) = 0$,

$$G(u,v) = 2\pi e^{-2\pi i\delta u} \int_0^{r_2} J_0(2\pi q\varrho)\varrho \, d\varrho =$$

$$= 2\pi r_2^2 e^{-2\pi i\delta u} \frac{J_1(2\pi q r_2)}{2\pi q r_2} . \tag{1.25}$$

This latter expression holds good, to be sure, only for the transparency function $g(\varrho, \zeta)$ as given by Equation (1.22), characteristic of an opaque circular disc.

Next—and this is essential—consider the fact that *the loss of light arising if the aperture centred at $(\delta, 0)$ begins to obscure that centre at $(0,0)$ can be defined as a cross-correlation between the two discs*—which is obviously zero if these do not overlap, and increases with an increase of their common intercept, weighted in accordance with the relative brightness of each element occulted (or the transparency of the respective area of the occulting star). Moreover, the measure of this cross-correlation is expressed by the *convolution integral* of the aperture function and eclipsing disc—i.e.,

$$L_1\alpha(r_1, r_2, \delta) = \int\int_{-\infty}^{+\infty} F(u,v)G(u,v)du\,dv , \tag{1.26}$$

where, by Equations (1.15) and (1.23), the integrand

$$F(u,v)G(u,v) = 2\pi r_2^2 L_1 e^{-2\pi i\delta u} \frac{J_1(2\pi q r_2)}{2\pi q r_2} \times$$

$$\times \sum_{n=0}^{N} C^{(n)} 2^\nu \Gamma(\nu) \frac{J_\nu(2\pi q r_1)}{(2\pi q r_1)^\nu} . \tag{1.27}$$

If, moreover, we remember that, in accordance with (1.5),

$$\left. \begin{array}{l} u = q \cos\phi, \quad v = q \sin\phi, \\ \\ du\,dv = q\,dq\,d\phi, \end{array} \right\} \tag{1.28}$$

it follows by use of (1.6)–(1.8) that

$$\int_{-\pi}^{\pi} e^{-2\pi i\delta q \cos\phi} d\phi = 2\pi J_0(2\pi q\delta). \tag{1.29}$$

Accordingly,

$$L_1\alpha(r_1, r_2, \delta) = \int_0^\infty \int_{-\pi}^{\pi} F(u,v)G(u,v)q\,dq\,d\phi =$$

$$= L_1(2\pi r_2)^2 \sum_{n=0}^{N} C^{(n)} 2^\nu \Gamma(\nu) \times$$

$$\times \int_0^\infty \frac{J_\nu(2\pi q r_1)}{(2\pi q r_1)^\nu} \frac{J_1(2\pi q r_2)}{2\pi q r_2} J_0(2\pi q\delta)q\,dq ; \tag{1.30}$$

and if we set, in acordance with Eq. (1.10) of Chapter II,

$$\alpha \;=\; \sum_{n=0}^{N} C^{(n)}\alpha_n^0 \,, \tag{1.31}$$

it follows by a comparison of (1.30) with (1.31) that the associated α-functions α_2^0 of order n are expressible as

$$\alpha_n^0(r_1, r_2, \delta) \;=\; (2\pi r_2)^2 2^\nu \Gamma(\nu) \times$$

$$\times \int_0^\infty \left\{ \frac{J_\nu(2\pi q r_1)}{(2\pi q r_1)^\nu} \frac{J_1(2\pi q r_2)}{2\pi q r_2} \right\} J_0(2\pi q \delta) q \, dq. \tag{1.32}$$

The right-hand side of Eq. (1.32) contains three parameters as multiplicative factors of the arguments of the Bessel functions occurring in it: namely, r_1, r_2 and δ. It is, however, easy to see that the functions α_n^0 depend on these parameters only through two *ratios* which can be formed of them. For if we set, say,

$$2\pi q r_2 \;\equiv\; x \,, \tag{1.33}$$

Equation (1.32) can be readily rewritten as

$$\alpha_n^0(h, k) \;=\; 2^\nu \Gamma(\nu) \int_0^\infty (kx)^{-\nu} J_\nu(kx) J_1(x) J_0(hx) dx \,, \tag{1.34}$$

where we have abbreviated

$$h \;=\; \frac{\delta}{r_2} \quad \text{and} \quad k \;=\; \frac{r_1}{r_2} \,. \tag{1.35}$$

It may be added that, for odd values of n (i.e., half-integral values of ν), the spherical Bessel functions $J_\nu(kx)$ become expressible in terms of trigonometric functions of kx.

Equation (1.34) represents the associated alpha-functions of index $m = 0$ and order $n = 1$ as *Hankel transforms* of zero order of the products of two Bessel functions of orders 1 and ν in curly brackets on the right-hand side of (1.34). This definition of α_n^0 is *more general* than that given by Eqs. (1.13) of Chapter II; for while those latter represent real quantities only if $\delta \leq r_1 + r_2$, and their explicit forms are, moreover, different for different types of eclipses (partial, total, annular), the definition (1.32) holds good for *any* type of eclipse and any value of r_1, r_2 or δ. In other words, the α_n^0's as given by Equation 1.32) represent real *and continuous non-negative functions of δ between minima as well as within eclipses of any type*; such that $\alpha_n^0(\delta)$—but not the function itself—may become discontinuous. For a proof of these facts the reader is referred, e.g., to Rice (1935) or Bailey (1936).

Moreover, Equation (3.34) can be readily generalized also to the case in which the light changes represented by it arise from eclipses by the discs which are *semi-transparent.* In order to prove this suppose (for the sake of argument) that the transparency of the occulting disc increases with the angle of foreshortening in the same manner as the limb-darkening of the eclipsed star—i.e., that the transparency function $g(\varrho, \zeta)$ in (1.21) varies as

$$g(\varrho, \zeta) = \begin{cases} [1 - (\varrho/r_2)^2]^\lambda, & \varrho \le r_2 ; \\ 0, & \varrho > r_2 ; \end{cases} \tag{1.36}$$

replacing (1.22) appropriate for opaque circular discs. If so, an exactly the same type of analysis as followed earlier in this section discloses that, in such a case, Eq. (1.25) for $G(u, v)$ is to be replaced by

$$G(u, v) = 2^{\lambda+1} \pi r_2^2 \Gamma(\lambda + 1) \frac{J_{\lambda+1}(2\pi q r_2)}{(2\pi q r_2)^{\lambda+1}} e^{-2\pi i \delta u} ; \tag{1.37}$$

and Eq. (1.34) for α_n^0, by

$$\alpha_n^0 = 2^{\lambda+\nu} \Gamma(\lambda + 1) \Gamma(\nu) k^{-\nu} \int_0^\infty x^{-\lambda-\nu} J_\nu(kx) J_{\lambda+1}(x) J_0(hx) dx , \tag{1.38}$$

which is of the same form as (1.34), except that the order of the second Bessel function behind the integral sign has been augmented by λ.

III.2   Hankel Transforms and Their Evaluation

Next, let us turn to the actual evaluation of the associated α-functions of the form α_n^0 as defined by the Hankel transform (1.34) in terms of the elements of the eclipses. In order to do so, we find it convenient to change over on the r.h.s. of (1.34) to a new variable y of integration, related with x by

$$x = \frac{r_2 y}{r_1 + r_2} , \tag{2.1}$$

so that

$$\alpha_n^0 = 2^\nu \Gamma(\nu) b \int_0^\infty (ay)^{-\nu} J_\nu(ay) J_1(by) J_0(cy) dy , \tag{2.2}$$

where

$$a = \frac{r_1}{r_1 + r_2} , \tag{2.3}$$

$$b = \frac{r_2}{r_1 + r_2} = 1 - a , \tag{2.4}$$

$$c = \frac{\delta}{r_1 + r_2} ; \tag{2.5}$$

such that, for any type of eclipse, $0 \le a \le 1, 0 \le b \le 1$—such that $a + b = 1$; while $1 \ge c \ge 0$ between the moment of first contact (when $\delta = r_1 + r_2$) and that of central eclipse ($\delta = 0$).

In order to evaluate the integrals on the r.h.s. of Eq. (2.2) we can avail ourselves of a theorem by Bailey (1936), asserting that

$$\int_0^\infty t^{\rho-1} J_\kappa(\alpha t) J_\lambda(\beta t) J_\mu(\gamma t) dt =$$

$$= \frac{2^{\rho-1}\alpha^\kappa\beta^\lambda\Gamma\left(\frac{\kappa+\lambda+\mu+\varrho}{2}\right)}{\gamma^{\kappa+\lambda+\rho}\Gamma(\kappa+1)\Gamma(\lambda+1)\Gamma\left(1-\frac{\kappa+\lambda-\mu+\varrho}{2}\right)} \times \tag{2.6}$$

$$\times F^{(4)}\left(\frac{\kappa+\lambda-\mu+\varrho}{2}, \frac{\kappa+\lambda+\mu+\varrho}{2}; \kappa+1, \lambda+1; \frac{\alpha^2}{\gamma^2}, \frac{\beta^2}{\gamma^2}\right),$$

where α, β, γ are positive quantities,

$$\varrho < \frac{5}{2}, \quad \kappa+\lambda+\mu+\varrho > 0 ; \tag{2.7}$$

and the Appell generalized hypergeometric series of the form

$$F^{(4)}(\mathbf{a}, \mathbf{b}; \mathbf{c}, \mathbf{c'}; x, y) \equiv \sum_{m=0}^\infty \sum_{n=0}^\infty \frac{(\mathbf{a})_{m+n}(\mathbf{b})_{m+n}}{m!n!(\mathbf{c})_m(\mathbf{c'})_n} x^m y^n \tag{2.8}$$

on the r.h.s. of Eq. (2.6) converges whenever

$$\gamma \geq \alpha + \beta . \tag{2.9}$$

Consider first the case in which

$$\alpha = a, \quad \beta = b, \quad \gamma = c. \tag{2.10}$$

If so, the condition (2.9) will be satisfied by (2.3)–(2.5) if

$$\delta \geq r_1 + r_2 \tag{2.11}$$

i.e., *outside eclipses* (the equality sign corresponding to the moment of first contact). If so, the constants

$$\kappa \equiv \nu, \quad \lambda \equiv 1, \quad \mu \equiv 1 - \nu \tag{2.12}$$

satisfy the conditions

$$\kappa + \lambda + \mu + \varrho = 2 > 0 \tag{2.13}$$

and

$$\varrho \equiv 1 - \nu = -\frac{1}{2}n < \frac{5}{2} ; \tag{2.14}$$

while

$$\Gamma\left(1 - \frac{\kappa+\lambda-\mu+\varrho}{2}\right) = \Gamma(0) = \infty . \tag{2.15}$$

Since the Appell series $F^{(4)}$ on the r.h.s. of Eq. (2.6) converges for (2.9), it follows from (2.2) and (2.6) that, outside eclipses,

$$\alpha_n^0(h \geq 1+k, k) = 0 ; \qquad (2.16)$$

not perhaps an unexpected result, but one which proves that the definition of α_n^0 in the form (1.1) or (2.2) holds good outside minima as well.

Consider now what happens when the eclipses become *total* or *annular*. The former occur whenever

$$r_2 \geq \delta + r_1 ; \qquad (2.17)$$

and if so, the condition (2.9) will be satisfied with

$$\alpha = a, \quad \beta = c, \quad \gamma = b \qquad (2.18)$$

and

$$\kappa \equiv \nu, \quad \lambda \equiv 0, \quad \mu \equiv 1, \quad \varrho \equiv 1 - \nu. \qquad (2.19)$$

If so, Equations (2.2) and (2.6) disclose that

$$\alpha_n^0(1 - k, k) = \frac{1}{\nu} = \frac{2}{n+2} , \qquad (2.20)$$

again in agreement with Eq. (1.15) of Chapter II.

Should, however, the eclipse become annular,

$$r_1 \geq \delta + r_2 , \qquad (2.21)$$

which conforms again to the condition (2.9) is

$$\alpha = b, \quad \beta = c, \quad \gamma = a \qquad (2.22)$$

and

$$\kappa \equiv 1, \quad \lambda \equiv 0, \quad \mu \equiv \nu, \quad \varrho \equiv 1 - \nu . \qquad (2.23)$$

In such a case, Eqs. (2.2) and (2.6) yield

$$\alpha_n^0 = (r_2/r_1)^2 F^{(4)}(1 - \nu, 1; 2, 1; r_2^2/r_1^2, \delta^2/r_1^2). \qquad (2.24)$$

For uniformly bright discs ($\nu = 1$) the r.h.s. of the foregoing equation reduces to $(r_2/r_1)^2$; and for even values of n (i.e., integral values of ν) it will become a polynomial; though for odd n's (i.e., half-integral ν's) the corresponding $F^{(4)}$ will remain an infinite series. If $\delta = 0$ (corresponding to the moment of central eclipse)

$$F^{(4)}(1 - \nu, 1; 2, 1; r_2^2/r_1^2, 0) = {}_2F_1(1 - \nu, 1, 2; r_2^2/r_1^2) ; \qquad (2.25)$$

and, therefore,

$$\alpha_n^0(0, r_2/r_1) = (r_2/r_1)^2 \, {}_2F_1(1 - \nu, 1; 2; r_2^2/r_1^2) = 1/\nu\{1 - (1 - r_2^2/r_1^2)^\nu\} \quad (2.26)$$

in agreement with Eq. (1.80) of Chapter II. Similarly, at the moment of internal tangency (when $\delta = r_1 - r_2$) Eq. (2.24) reduces to

$$\alpha_n^0 \left(\frac{r_1 - r_2}{r_1}, \frac{r_2}{r_1} \right) = \left(\frac{r_2}{r_1} \right)^2 F^{(4)} \left(1 - \nu, 1; 2, 1; \frac{r_2^2}{r_1^2}, \frac{(r_1 - r_2)^2}{r_1^2} \right) = \qquad (2.27)$$

$$= \frac{(2\nu)!}{\Gamma(\nu + 1)\Gamma(\nu + 2)} \left(\frac{r_2}{r_1} \right)^{\nu+1} {}_2F_1 \left(1 - \nu, \nu + 1; \nu + 2; \frac{r_2}{r_1} \right),$$

in agreement with Eq. (1.79) of Chapter II.

Should, however, the eclipse become *partial*, none of the three quantities r_1, r_2, δ (or a, b, c) can be equal to, or larger than, the sum of the two others. In consequence, the condition (2.7) can no longer be met, and Bailey's theorem (2.6) ceases to be applicable. In order to evaluate the integral on the r.h.s. of Equation (2.2) under these conditions, use can be made of an expansion due to Bateman (1905) of the form

$$J_\mu(y \cos \phi \cos \Phi) J_\nu(y \sin \phi \sin \Phi) = (2/y)(\cos \phi \cos \Phi)^\mu \times$$

$$\times (\sin \phi \sin \Phi)^\nu \sum_{j=0}^{\infty} (-1)^n (\mu + \nu + 2j + 1) \times$$

$$\times \frac{\Gamma(\mu + \nu + j + 1)\Gamma(\nu + j + 1)}{n!\Gamma(\mu + j + 1)\{\Gamma(\nu + 1)\}^2} \times$$

$$ {}_2F_1(-j, \mu + \nu + j + 1; \nu + 1; \sin^2 \phi) \times \qquad (2.28)$$

$$\times {}_2F_1(-j, \mu + \nu + j + 1; \nu + 1; \sin^2 \Phi) J_{\mu+\nu+2j+1}(y),$$

valid for all values of μ and ν other than those of negative integers.

Let us, in what follows, set $\mu \equiv 1$ and $\nu \equiv \nu$; while

$$\sin \phi \sin \Phi = a \quad \text{and} \quad \cos \phi \cos \Phi = b \qquad (2.29)$$

where a and b continue to be given by Equations (2.3) and (2.4). A solution of Equations (2.29) yields

$$\sin^2 \phi = \sin^2 \Phi = a, \qquad (2.30)$$

$$\cos^2 \phi = \cos^2 \Phi = b; \qquad (2.31)$$

so that the product of the first two Bessel functions on the r.h.s. of Eq. (2.2) can be expressed with the aid of Eq. (2.28) in the form of a summation

$$J_\nu(ay) J_1(by) = 2 \frac{a^\nu b}{y} \sum_{j=0}^{\infty} (-1)^j (\nu + 2j + 2) \frac{\Gamma(\nu + j + 1)\Gamma(\nu + j + 2)}{j!(j + 1)!} \times$$

$$\times \left\{ \frac{{}_2F_1(-k, k + \nu + 2; \nu + 1; a)}{\Gamma(\nu + 1)} \right\}^2 J_{\nu+2j+2}(y), \qquad (2.32)$$

where, it may be recalled, $b = 1 - a$. If, furthermore, we take advantage of the fact that

$$\int_0^\infty y^{-\nu-1} \, J_{\nu+2j+2}(y)J_0(cy)dy =$$

$$= \frac{2^{-\nu-1}j!}{\Gamma(\nu+j+2)} \, {}_2F_1(-\nu-j-1, j+1; 1; c^2), \qquad (2.33)$$

converging absolutely for $c^2 \leq 1$ for ν integral or fractional, a combination of Equations (2.2), (2.29) and (2.30) discloses that

$$\alpha_n^0 = (1-a)^2 \sum_{j=0}^\infty (-1)^j \frac{(\nu+2j+2)\Gamma(\nu+j+1)}{(j+1)!\nu\Gamma(\nu+1)} \times$$

$$\times \{{}_2F_1(-j, j+\nu+2; \nu+1; a)\}^2 \, {}_2F_1(-\nu-j-1, j+1; 1; c^2). (2.34)$$

We may note also that

$$(1-a){}_2F_1(-j, j+\nu+2; \nu+1; a) \equiv {}_2F_1(-j-1, j+\nu+1; \nu+1; a). \quad (2.35)$$

and

$$ {}_2F_1(-\nu-j-1, j+1; 1; c^2) \equiv (1-c^2)^{\nu+1} \, {}_2F_1(-j, \nu+j+2; \, c^2). \qquad (2.36)$$

Therefore, in terms of the Jacobi polynomials defined by

$${}_2F_1(-j, j+\alpha; \gamma; x) \equiv G_j(\alpha, \gamma; x) \qquad (2.37)$$

we can rewrite Equation (2.34) as

$$\alpha_n^0(a, c) = \frac{(1-c^2)^{\nu+1}}{\nu} \sum_{j=0}^\infty (-1)^j (\nu+2j+2) \frac{(\nu+1)_j}{(j+1)!} \times$$

$$\times \{G_{j+1}(\nu, \nu+1; a)\}^2 G_j(\nu+2, 1; c^2) =$$

$$= \frac{(1-c^2)^{\nu+1}}{(\nu)_2} \sum_{j=0}^\infty (j+1)(\nu+j+1)(\nu+2j+2) \times$$

$$\times \{bG_j(\nu+2, 2; b)\}^2 G_j(\nu+2, \nu+2; 1-c^2), \qquad (2.38)$$

where

$$(\nu)_j \equiv 1 \qquad \text{if } j = 0 \qquad (2.39)$$

$$\equiv \nu(\nu+1)(\nu+2)\cdots(\nu+j-1) \quad \text{if } j > 0$$

stand for the customary Pochhammer products, and where (cf., e.g., Slater, 1966; p.221) the squares of the Jacobi polynomials $G_j(\alpha, \gamma; x)$ can be expressed as Appell's generalized hypergeometric polynomials $F^{(4)}$ of the form (2.8), given by

$$\{G_j(\alpha, \gamma; x)\}^2 = \frac{(\gamma - \alpha - j)_j}{(\gamma)_j} F^{(4)}\{-j, \alpha + j; \gamma, 1 + \alpha - \gamma; x^2, (1 - x)^2\}. \quad (2.40)$$

The series on the right-hand side of Equation (2.38) represent the *most general* expansions for the associated α-functions of zero index and (not necessarily integral) order $n \geq -1$—valid for any type of eclipse; for our previous results obtained analytically in Section II-1 are valid only for integral values of n; and those represented by Equations (2.20) or (2.26) and (2.27) of this section, obtained by direct use of Bailey's theorem (2.6), represent likewise but particular cases of (2.38) for total or annular eclipses. Moreover, the arguments a, $b \equiv 1 - a$ and c defined by (2.3)–(2.5), have been constrained to remain within the interval $(0, 1)$ as shown on the accompanying Figure III.2. This latter figure discloses that, for

$$0 < a < \frac{1}{2}, \quad (2.41)$$

the eclipse giving rise to the light changes α_n^m is an occultation (signifying that $r_1 < r_2$); while if

$$\frac{1}{2} < a < 1, \quad (2.42)$$

the eclipse must be a transit (i.e., $r_1 > r_2$).

Moreover, the value of c as defined by (2.5) has been normalized so that

$$0 < c < 1. \quad (2.43)$$

If

$$1 > c > |2a - 1|, \quad (2.44)$$

the eclipse remains partial (regardless of whether $r_1 \gtrless r_2$). If, however,

$$c < 1 - 2a, \quad (2.45)$$

and $r_1 < r_2$, the eclipse becomes total; while if the converse is true and

$$c \leq 2a - 1, \quad (2.46)$$

it becomes annular. The equality sign corresponds to the moment of internal tangency, at which α_n^0 is given by Equation (2.27); while if $c = 0$, the eclipse becomes central, and α_n^0 given by Eq. (2.26). Lastly, the value of $c = 1$ corresponds to the moment of first contact of any type of eclipse; and if $c > 1$, there is no eclipse at all.

Should, on the other hand,

$$a = 0 \quad \text{and} \quad b = 1, \quad (2.47)$$

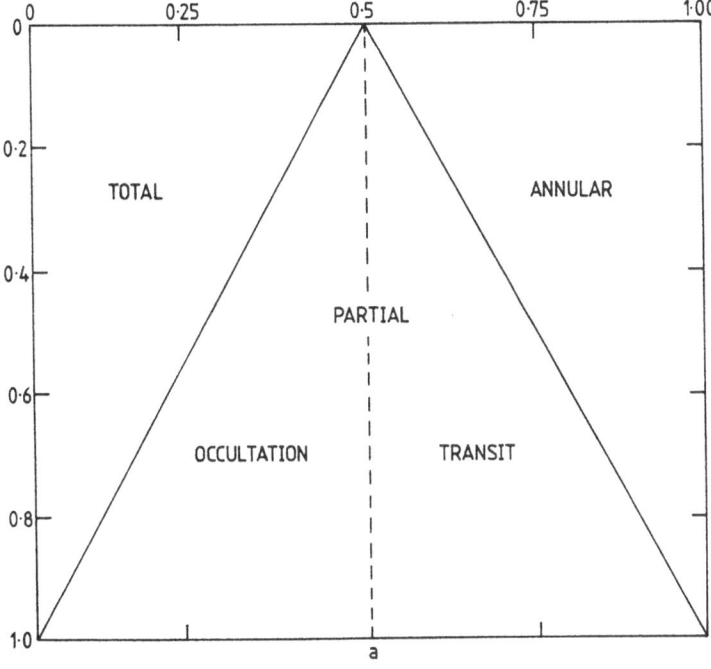

Figure III.2:

the occulting limb becomes a straight edge; and if so, the known asymptotic properties of Bessel functions enable us again to express the functions α_n^0 in a more elementary form. For consider what happens when both r_2 and δ tend to infinity, but in such a way that

$$\delta = r_2 + r_1 p \equiv r_2(1 + kp) . \tag{2.48}$$

As δ as well as r_2 grow large, it should be legitimate to resort to asymptotic expressions for Bessel functions to approximate

$$J_0(2\pi q\delta) \to \frac{\cos(2\pi q\delta - 45°)}{\pi\sqrt{q\delta}} \tag{2.49}$$

and

$$J_1(2\pi qr_2) \to \frac{\sin(2\pi qr_2 - 45°)}{\pi\sqrt{qr_2}} , \tag{2.50}$$

so that

$$\frac{J_2(2\pi qr_2)}{2\pi qr_2}J_0(2\pi q\delta) \to \frac{\cos 2\pi q(\delta + r_2) + \sin 2\pi qr_1 p}{\pi(2\pi qr_2)^2} . \tag{2.51}$$

If so, however, Equation (1.32) can be rewritten as

$$\pi\alpha_n^0 = -2^\nu\Gamma(\nu)\int_0^\infty y^{-\nu-1}\{\cos \Pi y + \sin py\}J_\nu\,dy , \tag{2.52}$$

where

$$y \equiv 2\pi q r_1 = kx , \tag{2.53}$$

and where we have abbreviated

$$\Pi = \frac{\delta + r_2}{r_1} \quad \text{and} \quad p = \frac{\delta - r_2}{r_1} . \tag{2.54}$$

Of the two integrals on the right-hand side of Eq. (2.52), it can be shown that

$$\lim \int_0^\infty y^{-\nu-1} J_\nu(y) \cos \Pi y \, dy = 0, \quad \Pi \to \infty \tag{2.55}$$

so that

$$\pi \alpha_n^0 = -2^\nu(\nu) \int_0^\infty y^{-\nu-1} J_\nu(y) \sin py \, dy; \tag{2.56}$$

rendering $\alpha_n^0(0, p)$ a Hankel transform, of ν-th order, of the function $y^{-\nu-1} \sin py$.

A repeated differentiation of (2.56) with respect to n discloses that, for eclipses by a straight edge,

$$\frac{\partial^2 \alpha_n^0}{\partial p^2} = 2^\nu \frac{\Gamma(\nu)}{\pi} \int_0^\infty y^{1-\nu} J_\nu(y) \sin py \, dy =$$

$$= \frac{2}{\sqrt{\pi}} \frac{\Gamma(\nu)}{\Gamma(\nu - \frac{1}{2})} p(1 - p^2)^{\nu-3/2} , \tag{2.57}$$

which on successive integration discloses that

$$\frac{\partial \alpha_n^0}{\partial p} = -\frac{\Gamma(\nu)}{\sqrt{\pi} \Gamma(\nu + \frac{1}{2})} (1 - p^2)^{\nu-1/2} \tag{2.58}$$

and

$$\alpha_n^0 = \frac{1}{2\nu} - \frac{\Gamma(\nu)}{\sqrt{\pi} \Gamma(\nu + \frac{1}{2})} \int_0^p (1 - p^2)^{\nu-1/2} dp =$$

$$= \frac{1}{2\nu} - \frac{p\Gamma(\nu)}{\sqrt{\pi} \Gamma(\nu + \frac{1}{2})} {}_2F_1\left(\frac{1}{2} - \nu, \frac{1}{2}; \frac{3}{2}; p^2\right) , \tag{2.59}$$

where the constant of integration has been adjusted so that, at the beginning of the eclipse, $\alpha_n^0(0, 1) = 0$. For $n = 0$ (corresponding to the eclipse of a uniformly bright disc by a straight occulting edge)

$$\pi \alpha_0^0 = \cos^{-1} \frac{s}{r_1} - \frac{s}{r_1} \sqrt{1 - \left(\frac{s}{r_1}\right)^2} ; \tag{2.60}$$

while if $n = 1$ (i.e., linear law of limb-darkening)

$$6\alpha_1^0 = 2 - 3\frac{s}{r_1} + \left(\frac{s}{r_1}\right)^3 . \tag{2.61}$$

These latter two equations should be no longer new to us; for if we remember that $r_1^2 - s^2 \equiv r_2^2 - (\delta - s)^2$, they represent only particular cases of Eqs. (1.23)–(1.24), Chapter II, for $m = 0$ and $n = 0, 1$.

III.3 Differential Properties of Hankel Transforms

In further development of our subject, let us turn to investigate the differential properties of the Hankel function employed in this section to express the associate alpha-functions by the equation (1.34), and compare them with those already established in the preceding section.

In order to do so, let us return to Eq. (1.34) and differentiate it with respect to the parameters r_1, r_2 and δ (involved in h and k) behind the integral sign. Since each one of these parameters occurs in the argument of only one of the Bessel functions constituting the integrand on the right-hand side of Equation (1.34), the use of the well-known recursion formulae

$$\frac{d}{dx}\left\{\frac{J_\nu(x)}{x^\nu}\right\} = -\frac{J_{\nu+1}(x)}{x^\nu} \tag{3.1}$$

and

$$\frac{2\nu}{x}J_\nu(x) = J_{\nu-1}(x) + J_{\nu+1}(x) \tag{3.2}$$

satisfied by Bessel functions of any order (integral or fractional), a partial differentiation with respect to $r_{1,2}$ and δ discloses that

$$r_1\frac{\partial\alpha_n^0}{\partial r_1} = -2^\nu\Gamma(\nu)k^{1-\nu}\int_0^\infty x^{1-\nu}J_{\nu+1}(kx)J_1(x)J_0(hx)dx , \tag{3.3}$$

$$r_2\frac{\partial\alpha_n^0}{\partial r_2} = 2^\nu\Gamma(\nu)k^{-\nu}\int_0^\infty x^{1-\nu}J_\nu(kx)J_0(x)J_0(hx)dx , \tag{3.4}$$

$$\delta\frac{\partial\alpha_n^0}{\partial\delta} = -2^\nu\Gamma(\nu)hk^{-\nu}\int_0^\infty x^{1-\nu}J_\nu(kx)J_1(hx)J_1(x)dx . \tag{3.5}$$

If we compare now the foregoing equations with (2.41)–(2.43) of Chapter II specifying the same derivatives in terms of the $I_{-1,n}^m$ integrals, it follows at once that for $0 \le m < 2$,

$$I_{-1,n}^m = 2^{\nu-1}\Gamma(\nu)k^\nu\int_0^\infty x^{1-\nu}J_\nu(kh)J_m(x)J_m(hx)dx =$$
$$= (2/b^2)^{\nu-1}a^\nu\Gamma(\nu)\int_0^\infty y^{1-\nu}J_\nu(ay)J_m(by)J_m(cy)dy , \tag{3.6}$$

where a, b, c and y are defined by Eqs. (2.1) and (2.3)–(2.5).

Since, moreover, by Bateman's expansion (2.28),

$$J_\nu(ay)J_m(by) = (2/y)a^\nu b^{-m}\sum_{j=0}^\infty(-1)^j(m+\nu+2j+1) \times$$
$$\times \frac{\Gamma(m+\nu+j+1)\Gamma(\nu+j+1)}{j!\Gamma(m+j+1)\{\Gamma(\nu+1)\}^2} \times$$
$$\times \{G_{m+j}(\nu+1-m, \nu+1, a)\}^2 J_{m+\nu+2j+1}(y) \tag{3.7}$$

while, by (2.33)

$$\int_0^\infty y^{-\nu}\, J_m(cy)J_{m+\nu+2j+1}(y)dy =$$

$$= \frac{\Gamma(m+j+1)c^m(1-c^2)^\nu}{2^\nu m!\Gamma(\nu+j+1)}\, G_j(m+\nu+1,\, m+1;\, c^2)\,,\quad (3.8)$$

it follows that

$$I_{-1,n}^m \;=\; \left(\frac{a}{b}\right)^{2\nu}\left(\frac{c}{b}\right)^m \frac{b^2(1-c^2)^\nu}{\nu} \sum_{j=0}^\infty \frac{(-1)^j}{j!}(m+\nu+2j+1)(\nu+1)_{m+j}\times$$

$$\times\, \{G_{m+j}(\nu+1-m,\, \nu+1,\, a)\}^2 G_j(m+\nu+1,\, m+1;\, c^2),\quad (3.9)$$

valid for $m = 0, 1$ and any value of n.

Moreover, by a combination of Equations (3.6) with (2.23) it follows at once that, for the indices $m = 0$ as well as 1,

$$\alpha_n^m \;=\; 2^\nu\Gamma(\nu)\int_0^\infty (kx)^{-\nu}J_{m+\nu}(kx)J_1(x)J_m(hx)dx \;=$$

$$= \;2^\nu\Gamma(\nu)b\int_0^\infty (ay)^{-\nu}J_{m+\nu}(ay)J_1(by)J_m(cy)dy,\quad (3.10)$$

which can likewise be represented (by the same method) by an expansion of the form

$$\alpha_n^m(a,c) \;=\; \frac{(ac)^m(1-c^2)^{\nu+1}}{(\nu)_{m+1}} \sum_{j=0}^\infty \frac{(-1)^j}{(j+1)!}(m+\nu+2j+2)\times$$

$$\times\, (j+1)_m(m+\nu+1)_{j+m}\{G_{j+1}(m+\nu,\, m+\nu+1,\, a)\}^2\times$$

$$\times\, G_j(m+\nu+2,\, m+1;\, c^2),\quad (3.11)$$

which for $m = 0$ reduces indeed to Eq. (2.38).

Furthermore, by a combination of the second half of Eq. (2.31) of Chapter II with (3.3) it follows at once that the functions $\Im_{-1,n}^0$ are likewise expressible as Hankel transforms of the form

$$\Im_{-1,n}^0 \;=\; 2^{\nu-1}\Gamma(\nu)\int_0^\infty (kx)^{1-\nu}J_{\nu+1}(kx)J_1(x)J_0(hx)dx\;;\quad (3.12)$$

and as the reader can easily verify, the recursion formula (2.31) established in Chapter II for $m = 0$ is nothing but a consequence of the recursion formula (3.2) for Bessel functions applied to the Hankel transform (1.34) for α_n^0.

The value of the integral on the r.h.s. of Eq. (3.12) can again be evaluated by an appeal to Bateman's expansion (3.7) and the integral formula (3.8): in doing so we similarly establish that

$$\Im^0_{-1,n} = \frac{a^2(1-c^2)^\nu}{\nu(\nu+1)} \sum_{j=0}^\infty \frac{(-1)^j}{j!}(\nu+2j+3)(\nu+2)_{j+1} \times$$

$$\times \{G_{j+1}(\nu+1,\nu+2;a)\}^2 G_{j+1}(\nu+1,c^2). \tag{3.13}$$

Next, let us turn to repeated differentiation of Eq. (1.32) with respect to δ. By setting

$$\delta^2 = r_1^2 - 2r_1 r_2 \cos \phi + r_2^2 \tag{3.14}$$

and making appeal to the addition theorem for Bessel functions we find that

$$J_0(h x) = J_0(x)J_0(kx) + 2\sum_{j=1}^\infty J_j(x)J_j(kx)\cos j\phi, \tag{3.15}$$

we can rewrite Eq. (1.34) as

$$a_n^0(h,k) = K_0^{(n)}(k) + 2\sum_{j=1}^\infty K_j^{(n)}(k)\cos j\phi, \tag{3.16}$$

where

$$\cos \phi = \frac{r_1^2 + r_2^2 - \delta^2}{2r_1 r_2} = \frac{1 - h^2 + k^2}{2k} \tag{3.17}$$

and

$$K_j^{(n)}(k) \equiv 2^\nu \Gamma(\nu) \int_0^\infty \frac{J_\nu(kx)}{(kx)^\nu} J_j(x)J_1(x)J_j(kx)dx. \tag{3.18}$$

The latter coefficients depend on k only; h occurring only in the argument of $j\phi$. Since, moreover,

$$\frac{\partial \phi}{\partial h^2} = \frac{1}{2k \sin \phi} \tag{3.19}$$

a differentiation of (3.16) with respect to h yields

$$\frac{\partial a_n^0}{\partial h^2} = \frac{1}{k \sin \phi} \sum_{j=1}^\infty K_j^{(n)}(k)\frac{\partial}{\partial \phi}(\cos j\phi) =$$

$$= -\frac{1}{k}\sum_{j=1}^\infty j K_j^{(n)}(k) U_j(\cos \phi), \tag{3.20}$$

where the $U_j(\cos \phi)$'s are Tchebyshev polynomials of the second kind. Moreover, an m-time differentiation of the preceding equation yields

$$\left(\frac{\partial}{\partial h^2}\right)^m a_n^0 = 2(2k \sin \phi)^{-m} \sum_{j=1}^\infty j^m K_j^{(n)}(k) R_e\{i^m e^{ij\phi}\}, \tag{3.21}$$

where $i \equiv \sqrt{-1}$.

In conclusion of the present chapter, devoted to a discussion of the formal properties of the eclipse function, let us return once more to the problem of a differential equation of the α_n^0's regarded as functions of h. In particular, we wish to prove that the associated α-functions zero index satisfy also a linear differential equation in h with *constant* coefficients—albeit at the expense of the fact that the order of this equation becomes infinite.

For if we abbreviate

$$h \frac{\partial}{\partial h} \equiv t , \tag{3.22}$$

a repeated application of this operator on (3.21) discloses that

$$t(t-2)(t-4) \cdots (t-2j)\alpha_n^0 =$$

$$= (-1)^{j+1} 2^\nu \Gamma(\nu) \int_0^\infty (kx)^{-\nu} J_\nu(kx) J_1(x) J_{j+1}(hx)(hx)^{j+1} dx. \tag{3.23}$$

Since, however, the Bessel functions $J_j(hx)$ are defined by the expansion

$$J_j(hx) = \sum_{i=0}^\infty \frac{(-1)^i (hx)^{2i+j}}{2^{2i+j}(i!)\Gamma(i+j+1)} \tag{3.24}$$

and (by one of Lommel's series; cf. Watson, 1945, p.141)

$$\sum_{j=0}^\infty \frac{J_j(hx)}{2^j(j!)}(hx)^j = 1 , \tag{3.25}$$

then by summing up an infinite number of equations of the type (3.23), divided each by $2^j j!$, we obtain

$$\left\{ 1 + \sum_{j=1}^\infty (-1)^j \frac{t(t-2)(t-4)\cdots(t-2j+2)}{2^j j!} \right\} \alpha_n^0 =$$

$$= 2^\nu \Gamma(\nu) \int_0^\infty \frac{J_\nu(kx)}{(kx)^\nu} J_j(x) \sum_{j=0}^\infty \frac{J_j(hx)}{2^j j!}(hx)^j dx =$$

$$= 2^\nu \Gamma(\nu) \int_0^\infty (kx)^{-\nu} J_\nu(kx) J_1(x) dx \tag{3.26}$$

by (3.25), which represents a linear differential equation for $\alpha_n^0(h)$ with constant absolute terms on the right-hand side.

The latter can, moreover, be readily evaluated for any type of eclipse in a closed form, for if $k < 1$ (corresponding to an occultation)

$$2^\nu \Gamma(\nu) \int_0^\infty (kx)^{-\nu} J_\nu(kx) J_1(x) dx = \frac{1}{\nu} , \tag{3.27}$$

which is equal to the maximum value of α_n^0 attained during totality; while if the eclipse is a transit $(k > 1)$

$$2^\nu \Gamma(\nu) \int_0^\infty (kx)^{-\nu} J_\nu(kx) J_1(x) dx = \frac{1 - (1 - k^{-2})^\nu}{\nu} \qquad (3.28)$$

signifies the loss of light at the moment of central eclipse.

If, therefore, we set

$$\alpha_n^0 = \begin{cases} \frac{1}{\nu} - f_n(h, k) & \cdots \text{ occultation} \\ \frac{1}{\nu}[1 - (1 - k^{-2})^\nu] - f_n(h, k) & \cdots \text{ transit} \end{cases} \qquad (3.29)$$

it follows that the differential equation (3.26) rewritten in terms of $f_n(h, k)$ will assume the form

$$\left\{ 1 + \sum_{j=1}^\infty (-1)^j \frac{t(t-2)(t-4)\cdots(t-2n+2)}{2^j \, j!} \right\} f_n = 0, \qquad (3.30)$$

which is homogeneous in the dependent variable, but not yet one with constant coefficients. The latter condition can, however, be attained if we change over from h to a new independent variable z defined by

$$h^2 = e^z; \qquad (3.31)$$

for if we abbreviate

$$\frac{\partial}{\partial z} \equiv D, \qquad (3.32)$$

the foregoing equation can be rewritten as

$$\left\{ 1 - D + \frac{1}{2!} D(D-1) - \frac{1}{3!} D(D-1)(D-2) - \cdots \right\} f_n \equiv \left\{ e^{-h^2 \frac{\partial}{\partial h^2}} \right\} f_n(h) = 0 \qquad (3.33)$$

Q.E.D.

III.4 Fourier Transforms and their Inversion

In conclusion of the present chapter, two problems remain still to be discussed: namely, the construction of a Fourier transform of the loss of light, during eclipses, of spherical stars arbitrarily darkened at the limb; and, as a by-product of such a transform, an expansion of this loss in terms of the cosines of integral multiples of the *phase angle* ψ (for its definition, cf. Eqs. (1.23) of Chapter IV); with coefficients given as explicit functions of the elements of the eclipse. In order to evaluate the Fourier transforms of the light curves of eclipsing variables, we depart from the Fourier theorem in the form

$$F(\nu) = \int_{-c}^{c} f(\psi) e^{-2\pi i \nu \psi} d\psi, \qquad (4.1)$$

$$f(\psi) = \int_{-\infty}^{\infty} F(\nu) e^{2\pi i \nu \psi} d\nu, \qquad (4.2)$$

where $f(\psi)$ denotes an 'input function'—such as represented, e.g., by the light changes of an eclipsing variable as functions of the phase angle ψ— and $F(\nu)$, its Fourier transform regarded as a function of frequency ν. The limits $\pm c$ of integration on the right-hand side of Equation (4.1) can be finite or infinite— depending (in effect) on the range of ψ within which the function $f(\psi)$ is different from zero. For example, if $f(\psi)$ represents a loss of light arising from the eclipses of spherical stars, the quantities $\pm c$ may be identified with the moments of the first and last contact ψ_1 of the eclipses; for since $f(\psi) = 0$ for $\psi > c$, the value of the integral on the r.h.s. of (4.1) would clearly be unaffected by any further extension of the limits.

As is well known, even for real functions $f(\psi)$ their transform $F(\nu)$ will generally be a complex quantity. In order to separate its real and imaginary part, let us consider the identity

$$f(\psi) = \frac{1}{2}[f(\psi) + f(-\psi)] + \frac{1}{2}[f(\psi) - f(-\psi)] , \tag{4.3}$$

by virtue of which

$$\int_{-c}^{c} \frac{1}{2}[f(\psi) + f(-\psi)]e^{ih\psi}\, d\psi =$$

$$= \frac{1}{2}\int_{0}^{c} [f(\psi) + f(-\psi)](e^{ih\psi} + e^{-ih\psi})\, d\psi =$$

$$= \int_{0}^{c} [f(\psi) + f(-\psi)] \cos h\psi\, d\psi ; \tag{4.4}$$

and, similarly,

$$\int_{0}^{c} \frac{1}{2}[f(\psi) - f(-\psi)]e^{-ih\psi}\, d\psi =$$

$$= \frac{1}{2}\int_{0}^{c} [f(\psi) - f(-\psi)](e^{-ih\psi} - e^{ih\psi})\, d\psi =$$

$$= -i\int_{0}^{c} [f(\psi) - f(-\psi)] \sin h\psi\, d\psi , \tag{4.5}$$

where we have abbreviated

$$h \equiv 2\pi\nu . \tag{4.6}$$

In consequence, the Fourier transform $F(\nu)$ as defined by Equation (4.1) can obviously be expressed as

$$F(\nu) = F_1(\nu) - iF_2(\nu) , \tag{4.7}$$

where

$$F_1(\nu) = \int_{0}^{c} [f(\psi) + f(-\psi)] \cos(2\pi\nu\psi)\, d\psi \tag{4.8}$$

and

$$F_2(\nu) = \int_0^c [f(\psi) - f(-\psi)] \sin(2\pi\nu\psi) \, d\psi \tag{4.9}$$

are—for real functions $f(\psi)$—both real quantities.

Equation (4.2) of the Fourier theorem discloses then that the original input function $f(\psi)$ can be synthesized from its transform $F(\nu)$ by

$$
\begin{aligned}
f(\psi) &= \int_{-\infty}^{\infty} [F_1(\nu) - iF_2(\nu)] e^{2\pi i\nu\psi} \, d\nu = \\
&= \int_0^{\infty} F_1(\nu)(e^{2\pi i\nu\psi} + e^{-2\pi i\nu\psi}) \, d\nu - \\
&\quad - i \int_0^{\infty} F_2(\nu)(e^{2\pi i\nu\psi} - e^{-2\pi i\nu\psi}) \, d\nu = \\
&= 2 \int_0^{\infty} [F_1(\nu) \cos 2\pi\nu\psi + F_2(\nu) \sin 2\pi\nu\psi] \, d\nu = \\
&= 2 \int_0^{\infty} \sqrt{F_1^2 + F_2^2} \cos\left(2\pi\nu\psi - \tan^{-1} \frac{F_2}{F_1}\right) \, d\nu .
\end{aligned}
\tag{4.10}
$$

Let us apply now the foregoing formulae to an evaluation of the Fourier transform for the case when the input funtion

$$f(\psi) \equiv 1 - l \equiv f(-\psi) \tag{4.11}$$

represents the loss of light

$$f(\psi) \equiv 1 - l = L_1\alpha = L_1 \sum_{n=0}^{N} C^{(n)}\alpha_n^0 , \tag{4.12}$$

with symmetrical eclipses, where the coefficients $C^{(n)}$ continue to be given by Eqs. (1.11)–(1.12) of Chapter II; and the fractional loss α_n^0 of light, by (2.38).

If so, its Fourier transform (4.8) will reduce to

$$F_1(\nu) = 2 \int_0^{\psi_1} (1 - l) \cos h\psi \, d\psi , \tag{4.13}$$

where (cf., e.g., Oberhettinger, 1973), for an arbitrary value of $h > 0$,

$$\cos h\psi = \frac{\sin \pi h}{\pi h} \left\{ 1 + 2 \sum_{m=1}^{\infty} \frac{(-1)^m h^2}{h^2 - m^2} \cos m\psi \right\} \tag{4.14}$$

for $-\pi < \psi < \pi$, which for integral values of $h \equiv m$ reduces to

$$\cos m\psi = {}_2F_1\left(-\frac{m}{2}, \frac{m}{2}; \frac{1}{2}; \sin^2 \psi\right) . \tag{4.15}$$

Let us, furthermore, replace the time-variable ψ by a new variable u defined by the equation

$$u \equiv \frac{\delta^2 - \delta_0^2}{\delta_1^2 - \delta_0^2} = \left(\frac{\sin \psi}{\sin \psi_1}\right)^2 ; \tag{4.16}$$

a differentiation of which discloses that

$$d\psi = \frac{\sin \psi_1 \, du}{2\sqrt{u(1 - u \sin^2 \psi_1)}} . \tag{4.17}$$

By a combination of Eqs. (4.14)–(4.17), Equation (4.13) can be rewritten as

$$F_1(\nu) = L_1 \sin \psi_1 \sum_{n=0}^{N} C^{(n)} \frac{\sin \pi h}{\pi h} \int_0^1 \Bigg\{ 1 +$$

$$+ 2 \sum_{m=1}^{\infty} \frac{(-1)^m h^2}{h^2 - m^2} \, {}_2F_1\left(-\frac{m}{2}, \frac{m}{2}; \frac{1}{2}; u \sin^2 \psi_1\right) \Bigg\} \times$$

$$\times \frac{\alpha_n^0 \, du}{\sqrt{u(1 - u \sin^2 \psi_1)}} , \tag{4.18}$$

in which the argument $1 - c^2$ in α_n^0 can, by virtue of (4.16), be rewritten in terms of u as

$$1 - c^2 = (1 - c_0^2)(1 - u) , \tag{4.19}$$

where (in accordance with Eq. 2.5) $c = \delta/\delta_1$ and $c_0 = \delta_0/\delta_1$.
Since, moreover,

$$(1 - u \sin^2 \psi_1)^{-1/2} \, {}_2F_1\left(-\frac{m}{2}, \frac{m}{2}; \frac{1}{2}; u \sin^2 \psi_1\right) =$$

$$= {}_2F_1\left(\frac{1 - m}{2}, \frac{1 + m}{2}; \frac{1}{2}; u \sin^2 \psi_1\right) , \tag{4.20}$$

the integrals on the r.h.s. of Eq. (4.18) can be expressed in terms of the form

$$\int_0^1 u^{-1/2}(1 - u)^j \, {}_2F_1\left(\frac{1 - m}{2}, \frac{1 + m}{2}; \frac{1}{2}; u \sin^2 \psi_1\right) du =$$

$$= B\left(\frac{1}{2}, j + 1\right) {}_2F_1\left(\frac{1 - m}{2}, \frac{1 + m}{2}; j + \frac{3}{2}; \sin^2 \psi_1\right) ; \tag{4.21}$$

and if so, the Fourier transform $F_1(\nu)$ of argument $\nu \equiv h/2\pi$, on insertion for α_n^0 from (2.38), can be expressed as

$$F_1(\nu) = L_1 \sin \psi_1 \left(\frac{\sin \pi h}{\pi h}\right) \sum_{n=0}^{N} C^{(n)} \times$$

$$\times \sum_{j=0}^{\infty} \left(\frac{n+6}{2} + 2j \right) \left[bG_j \left(\frac{n+6}{2}, 2; b \right) \right]^2 \sum_{i=0}^{j} \frac{(-1)^i (j+1)!}{i!(j-i)!} \times$$

$$\times \left(\frac{n+4}{2} + j \right)_{i+1} (1-c^2)^{\frac{n+4}{2}+1} \left[\frac{\Gamma\left(\frac{1}{2}\right) \Gamma\left(\frac{n+2}{2}\right)}{\Gamma\left(\frac{n+7}{2} + i\right)} \right] \times$$

$$\times \left[{}_2F_1 \left(\frac{1}{2}, \frac{1}{2}; \frac{n+7}{2} + i; \sin^2 \psi_1 \right) + \right. \tag{4.22}$$

$$\left. + 2h^2 \sum_{n=1}^{\infty} \frac{(-1)^m}{h^2 - m^2} \, {}_2F_1 \left(\frac{1-m}{2}, \frac{1+m}{2}; \frac{n+7}{2} + i; \sin^2 \psi_1 \right) \right]$$

The establishment of the foregoing result has completed only one-half of the task set forth in this section; for our remaining task should be to synthesize the transform $F_1(\nu)$ to reproduce the loss of light established already in the preceding section. Although the result to be obtained is already known to us, the aim of the Fourier synthesis still to be performed should enable us to express the function α_n^0 in terms of the cosines of integral multiples of the phase angle ψ rather than in ascending powers of $1 - c^2$ previously used.

In order to do so, let us recall that this can be accomplished simply by an inversion of the continuous frequency-spectrum $F_1(\nu)$, by a resort to the second half of the Fourier theorem represented by equation (4.2)—or, more specifically, by evaluating an integral of the product $F_1(\nu) \cos 2\pi\nu\psi$ with respect to the frequency ν over the entire half-plane—as

$$f(\psi) \equiv L_1 \sum_{n=0}^{\infty} C^{(n)} \alpha_n^0 = 2 \int_0^{\infty} F_1(\nu) \cos 2\pi\nu\psi \, d\nu \equiv$$

$$\equiv \frac{1}{\pi} \int_0^{\infty} F_1(h) \cos h\psi \, dh , \tag{4.23}$$

where $F_1(h)$ is already known to us from Equation (4.22), and $\cos h\psi$ continues to be given by Eq. (4.14).

The task we face is indeed simple; for the only terms in (4.22) which depend on our variable h of integration are of the form $(h \sin \pi h)/(h^2 - m^2)$; and since

$$\int_0^{\infty} \frac{\sin \pi h \cos h\psi}{h^2 - m^2} dh^2 = \pi (-1)^m \cos m\psi \tag{4.24}$$

for $\pi > \psi > -\pi$ and $m \geq 0$, by equating the terms on both sides of Eq. (4.23) factored by equal coefficients $C^{(n)}$ we find that

$$\alpha_n^0 = \frac{1}{2} B_0^{(n)} + \sum_{m=1}^{\infty} B_m^{(n)} \cos m\psi , \tag{4.25}$$

where

$$B_m^{(n)} = \Gamma\left(\frac{1}{2}\right)\Gamma\left(\frac{n+2}{2}\right) \sin \psi_1 \sum_{j=0}^{\infty} \left(\frac{n+6}{2} + 2j\right) \times$$

$$\times \left\{ bG_j\left(\frac{n+6}{2}, 2; b\right)\right\}^2 \sum_{i=0}^{\infty} \frac{(-1)^i (j+1)!}{i!\,(j-1)!} \times$$

$$\times \left(\frac{n+4}{2} + j\right)_{i+1} \frac{(1-c_0^2)^{\frac{n+4}{2}+i}}{\Gamma\left(\frac{n+7}{2}+i\right)} \times$$

$$_2F_1\left(\frac{1-m}{2}, \frac{1+m}{2}; \frac{n+7}{2} + i; \sin^2 \psi_1\right) \tag{4.26}$$

as a Fourier expansion for α_n^0 in terms of $\cos m\psi$ $(m = 0, 1, 2, \ldots)$, valid whenever $\alpha_n^0(\psi) = \alpha_n^0(-\psi)$; which can be rewritten as

$$B_m^{(n)} = \frac{\sqrt{\pi}\,\Gamma(\nu)}{\Gamma(\nu + \frac{5}{2})}(1-c_0^2)^{\nu+1}b^2 \sin \psi_1 \times$$

$$\times \sum_{j=0}^{\infty} \frac{(-1)^j}{j!}(\nu+j)_j(\nu+2j+2)F^{(4)}\left\{-j, \nu+j+2;\right.$$

$$2, \nu+1; b^2, 1-b^2\right\} F^{(3)}\left\{-j, \frac{1-m}{2}; \nu+j+2,\right.$$

$$\left. \frac{1+m}{2}; \nu+\frac{5}{2}; 1-c_0^2, \sin^2 \psi_1 \right\} \tag{4.27}$$

as a series of the products of Appell's generalized hypergeometric series $F^{(3)}$ and $F^{(4)}$ of two (respective) variables; and where we have abbreviated again $n + 2 \equiv 2\nu$.

Should this latter condition not be met—as it will not if (in general) the relative orbit of the two components are eccentric; if one (both) components of the system exhibit free oscillations; or for any other cause—we know already from Eq. (4.7) that the Fourier transform of asymmetric light changes becomes a *complex* quantity. Its imaginary part $F_2(\nu)$ can, in turn, be evaluated in much the same way as we did for $F_1(\nu)$, with the aid of the formulae (cf. again Oberhettinger, op.cit.)

$$\sin h\psi = 2\frac{\sin \pi h}{\pi} \sum_{m=1}^{\infty} \frac{(-1)^m m}{h^2 - m^2} \sin m\psi, \tag{4.28}$$

$$\sin m\psi = m \sin \psi \,_2F_1\left(\frac{1-m}{2}, \frac{1+m}{2}; \frac{3}{2}; \sin^2 \psi\right); \tag{4.29}$$

but the details of its actual evaluation can be left as an exercise for the interested reader.

III.5 Radial Velocities: Rotational Effect

The principal subject of this chapter so far has been to express mathematically the fractional loss of light during eclipses in binary systems as cross-correlations of circular apertures—one representing the star undergoing eclipse, and the other the eclipsing disk—the former arbitrarily darkened at the limb, and the latter of arbitrary transparency (semi-transparent if the eclipse is atmospheric). Outside eclipses, this cross-correlation is zero; while within eclipses it can be expressed in terms of Hankel transforms of zero order. In the next Chapter IV the same idea will then be extended to close binary systems, whose light varies continuously in the course of the entire orbital cycle by virtue of their distortion.

The tell-tale variations of light of close binary systems do not, to be sure, represent the only hallmark of their nature; as is well known, another is represented by the variation of *radial velocity* exhibited by the rotating components of such systems and observable by spectroscopic means. If the latter can be regarded as spheres which do not eclipse each other, their light (apart from the reflection effect) can be regarded as constant. Not so, however, their radial velocities; for even if the centres of light of the two components project themselves constantly on their centres of mass, their velocities—including their components in the line of sight—continue to be governed by their Keplerian motion.

Should, however, the rotation and tides in close binary systems cause the equilibrium surfaces of their components to deviate from spherical form, their centres of light and mass will no longer coincide in projection with those of their centres of mass; and, as a result, the radial velocities exhibited by them for distant observers will no longer be identical with those of their Keplerian motions (the same would be true also for spherical stars of finite size, because of their mutual irradiation); during minima, such phenomena would be further exaggerated. While the main contents of the present tract will be concerned with the light changes of close binary systems, the main aim of the concluding section of this chapter will be to demonstrate that although the principal phenomena characterizing close binary systems—namely, their variations of light and of radial velocities during eclipses—are physically quite different and observable by very different means, their effects can be mathematically described by functions of very much the same kind.

In order to show this to be the case, let V_{rot} stand for the radial component of the velocity-vector at any surface point of a distorted star rotating about the axis perpendicular to the orbital plane with an angular velocity ω_z, the "rotational effect" δV on the observed radial velocity V of the star as a whole should be represented by the ratio

$$\delta V \equiv \frac{\int V_{rot}\, d\ell}{\int d\ell}, \tag{5.1}$$

where the denominator $\int d\ell$ is identical with the luminosity of the rotating component (or, in eclipsing systems, for a visible fraction L thereof), and

$$V_{\text{rot}} = \omega_z(n_1 x - n_2 y), \tag{5.2}$$

in which the direction cosines $n_{1,2}$ are given by Eqs. (1.14)–(1.15) of Chapter IV, and the coordinates x, y are identical with those introduced in Section 1 of that chapter (for the more general case of three-axial rotation, the foregoing equation (5.2) can be easily generalized to the form given by Eq. (4.13), Chapter VII of Kopal, 1989).

In order to evaluate the numerator on the right-hand side of Eq. (5.2), let us return to Eq. (1.1) of this chapter, in which

$$f(x,y) \equiv f(r) V_{\text{rot}} \equiv$$

$$\equiv H\{1 - u_1 - u_2 - u_3 - \cdots + \sum_{n=0}^{N} u_n \cos^n \gamma\} \times$$

$$\times \omega_z r(n_1 \cos \theta - n_2 \sin \theta) \tag{5.3}$$

and $\cos \gamma = \sqrt{r_1^2 - r^2}/r_1$; so that Eq. (1.4) should be generalized to

$$f(u,v) = \omega_z H \sum_{n=0}^{N} \frac{C^{(n)}}{r_1^n} \left\{ n_1 \int_0^\infty \int_{-\pi}^{\pi} (r_1^2 - r^2)^{n/2} \times \right.$$

$$\times e^{-2\pi i q r \cos(\theta - \phi)} r^2 dr \cos \theta \, d\theta - \tag{5.4}$$

$$\left. - n_2 \int_0^\infty \int_{-\pi}^{\pi} (r_1^2 - r^2)^{n/2} e^{-2\pi i q r \cos(\theta - \phi)} r^2 dr \sin \theta \, d\theta \right\}.$$

Let us proceed—as in Section 1—to evaluate the foregoing integrals by a resort to the Jacobi expansion (1.6). The only non-vanishing terms which replace (1.8) are

$$\int_{-\pi}^{\pi} \sin(2n+1) \left(\frac{\pi}{2} + \theta - \phi\right) \left\{ \begin{array}{c} \cos \theta \\ \sin \theta \end{array} \right\} d\theta = \pi \left\{ \begin{array}{c} \cos \phi \\ \sin \phi \end{array} \right\}, \tag{5.5}$$

so that

$$\int_{-\pi}^{\pi} e^{-2\pi i q r \cos(\theta - \phi)} \left\{ \begin{array}{c} \cos \theta \\ \sin \theta \end{array} \right\} d\theta = -2\pi i \, J_1(2\pi q r) \left\{ \begin{array}{c} \cos \phi \\ \sin \phi \end{array} \right\} \tag{5.6}$$

and, accordingly,

$$F(q,\phi) = -2\pi i \omega_z H \sum_{n=0}^{N} \frac{c^{(n)}}{r_1^n} (n_1 \cos \phi - $$

$$- n_2 \sin \phi) \int_0^\infty (r_1^2 - r^2)^{n/2} J_1(2\pi q r) r^2 dr; \tag{5.7}$$

where the upper limit of integration can be reduced again from ∞ to r_1.

Now (cf. Erdélyi *et al*, 1954; 26(33))

$$\int_0^{r_1} (r_1^2 - r^2)^{n/2} J_1(2\pi qr) r^2 dr = 2^{\nu-1} r_1^{2\nu+1} \Gamma(\nu)(2\pi q r_1)^{-\nu} J_{\nu+1}(2\pi q r_1), \quad (5.8)$$

where $\nu = \frac{1}{2}(n+2)$; while the Fourier transform $G(u,v)$ of the eclipsing aperture continues to be given by Equaton (1.25). In evaluating the convolution of $F(u,v)G(u,v)$ as defined by Eqs. (1.30)–(1.32) with the aid of the preceding results, note that

$$\int_{-\pi}^{\pi} e^{-2\pi i \delta q \cos \phi} \left\{ \begin{array}{c} \cos \phi \\ \sin \phi \end{array} \right\} d\phi = -2\pi i \left\{ \begin{array}{c} 1 \\ 0 \end{array} \right\} J_1(2\pi \delta q). \quad (5.9)$$

Accordingly, the rotational factor

$$L_{\mathfrak{C}} \, \delta V \;=\; -\omega_z n_1 r_1 \sum_{n=0}^{N} C^{(n)} 2^{\nu} \Gamma(\nu) \times$$

$$\times \int_0^{\infty} \frac{J_{\nu+1}(kx)}{(kx)^{\nu}} J_1(x) J_1(hx) dx, \quad (5.10)$$

where the parameters h and k continue to be given by Eqs. (1.35); and which with the aid of (3.10) can be rewritten as

$$L_{\mathfrak{C}} \, \delta V \;=\; -\omega_z n_1 r_1 \sum_{n=0}^{N} C^{(n)} \alpha_n^1(h,k) \quad (5.11)$$

in which $\alpha_n^1(h,k)$ stands for the associated alpha-function α_n^m of order n and index $m = 1$ in accordance with Eq. (3.11).

Moreover, when we turn to investigate the effects, on the radial velocity, of distorted components of close binary systems (cf. Kopal, 1945), the same feature transpires from our analysis: namely, that the difference between the effects on light, and radial velocity, curves of distorted binary systems arises from the fact that *the indices of all associated α-functions needed to describe such an effect should be increased by unity*.

III.6   Bibliographical Notes

A formulation of all types of the eclipse functions in terms of the Hankel transforms, as given in Section III-1, is due to Kopal (1977b,c) and subsequent literature quoted in the text. However, only the first steps have been taken so far to express the respective eclipse functions in terms of other types of integral transforms—particularly the Laplace transforms. The existence of relations between the Hankel and Laplace transforms were pointed out by Tricomi in 1935; and applied to our problems by Kopal (1983); but this subject has not been developed in this book.

The role of Fourier transforms in this connection, discussed in Sec. III-4, was investigated first by Kopal (1977a; cf. also Sec. I-4 of 1979); and also by Kopal and Yamasaki (1980); for extensive tabulations of certain such functions cf. Kitamura (1967).

For a fuller discussion of integrals of the form (1.38) for an arbitrary transparency parameter $\lambda > 0$ on the r.h.s. of that equation cf. Srivastava and Kopal (1989).

The "rotational effect" within eclipses—detected observationally first by Schlesinger (1909) in δ Librae, and subsequently by McLaughlin (1924) in Algol and Rossiter (1924) in β Lyrae—was theoretically analyzed (for spherical stars) by Petrie (1938) and, more fully (in terms of the associated α-functions) by Kopal (1942b); to whom our present Equation (5.11) is due. An extension of this theory to systems consisting of distorted components is likewise due to Kopal (1945); for its summary see Chapter V of Kopal (1959) or Chapter VII of Kopal (1989).

Chapter IV

THEORETICAL PHOTOMETRY OF DISTORTED ECLIPSING SYSTEMS

In the preceding chapter of this book an outline has been given of the mathematical theory of eclipsing binary systems as long as their components (characterized by an arbitrary distribution of brightness on their apparent discs) can be regarded as spherical, and would appear in projection on the celestial sphere as circular discs. The question is, however, bound to be raised as to the extent to which such a model can be regarded as a satisfactory representation of *close* binaries actually encountered in the sky. This would, in principle, be legitimate only in the absence of forces which distort the shape of both components and cause their equipotential surfaces to depart from spheres: these are, however, ever present in the form of *axial rotation* of the components, and of their mutual *tidal action*. A theory of the effects produced by such forces has recently been summarized by the present writer in another volume (cf. Kopal, 1989), to which the reader is referred for fuller details that need not be repeated in this place.

The essential feature of this problem rests on the fact that the amplitude of the photometric effects in close eclipsing systems, caused by the rotational as well as tidal distortion, grows rapidly with increasing proximity of their components. In many known eclipsing systems, the components are, to be sure, separated widely enough—and, consequently, their distortion is sufficiently small—that (within the limits of observational errors) they can be regarded as spheres; and for such systems a theory of the light changes as given in the preceding chapter should indeed remain adequate. However, if the components are closer to each other—and (apart from the far greater intrinsic interest in such binaries) observational selection strongly favours discovery of close systems—a situation arises requiring a new approach. For a significant feature of "close" eclipsing systems is the fact that *observable light changes are no longer confined only to the times of the minima, but extend over the entire cycle*. This is due to two reasons. First, since the components of close binaries possess, in general, the form of distorted ellipsoids with longest axes oriented constantly in the direction of the line joining their centres, their apparent cross-sections—and, therefore, the light—exposed to the observer should vary continuously in the course of a revolution ("ellipticity effect"). Secondly, it is also inevitable in close binaries that a fraction of the radiation of each component will be intercepted by its mate, to be absorbed and re-emitted (or scattered) in all directions—including that of the line of sight. The amount of light so "reflected" by each component towards the observer will again vary with the phase ("reflection effect").

The changes of light arising from the ellipticity and reflection are independent of, and supplementary to, the light changes which arise if our binary system happens to be an eclipsing variable; and the magnitude of each increases with increasing proximity of the components. The way in which these can be separated will be discussed later in Chapter VI. However, before we come to do so, we must investigate first the nature of the theoretical light curves of close eclipsing systems—between minima as well as within eclipses—and the aim of the present chapter will be to carry out such an investigation to serve as a basis for all subsequent work.

IV.1   Theoretical Light Changes: Transformation of Coordinates

In order to investigate the light changes exhibited by axial rotation of distorted components of close binary syhstems, let us begin by introducing four systems of rectangular coordinates in which all aspects of such phenomena can be appropriately treated: namely,

(1) The (unprimed) axes XYZ will represent a system of *inertial* coordinates ("space axes") of directions fixed in space in such a way that the XY-plane coincides with the *invariable plane of the system*; while the Z-axis is perpendicular to it.

(2) The singly-primed axes $X'Y'Z'$ will stand for a system of rectangular coordinates *rotating* with the body ("body-axes"), defined so that $X'Y'$-plane represents the (instantaneous) *equator* of the rotating star, inclined by an angle θ to the inertial XY-plane and intersecting it at the angle ϕ (see Figure IV.1).

(3) The doubly-primed axes $X''Y''Z''$ will hereafter represent a system of *revolving* rectangular coordinates, in which the X''-axis is constantly coincident with the radius-vector between the origin and the centre of mass of the revolving star, and $Z'' = 0$ represents the (instantaneous) position of the orbital plane.

(4) The triply–primed axes $X'''Y'''Z'''$ will, lastly, represent another system of moving coordinates, in which the Z'''-axis coincides with the *line-of-sight*, and the X'''-axis is made to coincide with the projection of the X''-axis on the plane tangent to the celestial sphere passing through the origin of coordinates (i.e., $Z''' = 0$).

As is well known, a transformation of coordinates from the inertial (space) to the rotating (body) axes is governed by the matrix equation

$$
\left\{ \begin{array}{c} x \\ y \\ z \end{array} \right\} = \left\{ \begin{array}{ccc} a'_{11} & a'_{12} & a'_{13} \\ a'_{21} & a'_{22} & a'_{23} \\ a'_{31} & a'_{32} & a'_{33} \end{array} \right\} \left\{ \begin{array}{c} x' \\ y' \\ z' \end{array} \right\} \tag{1.1}
$$

(and its inverse follows from a transposition of the columns and rows of the square

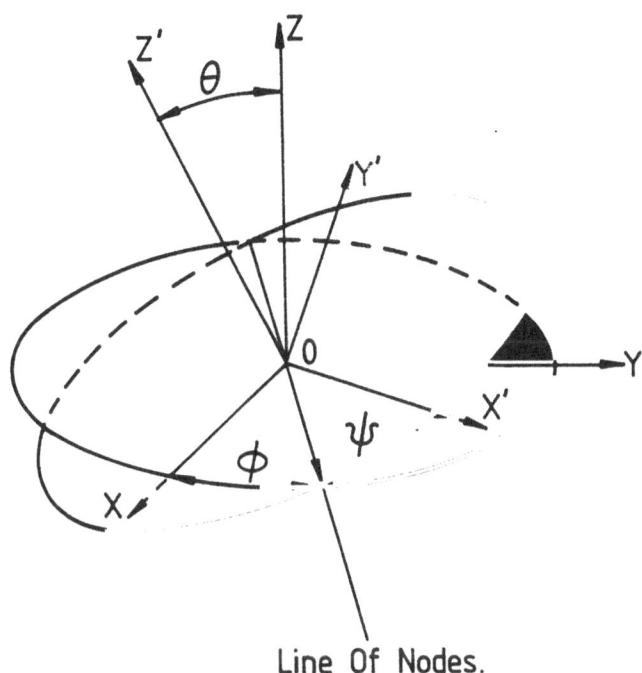

Line Of Nodes.

Figure IV.1: Definition of the Eulerian angles.

matrix on the right-hand side), where the direction cosines

$$
\left.\begin{aligned}
a'_{11} &= \cos \varphi \cos \phi - \sin \varphi \sin \phi \cos \theta, \\
a'_{12} &= -\sin \varphi \cos \phi - \cos \varphi \sin \phi \cos \theta, \\
a'_{13} &= \sin \phi \sin \theta;
\end{aligned}\right\} \tag{1.2}
$$

$$
\left.\begin{aligned}
a'_{21} &= \cos \varphi \sin \phi + \sin \varphi \cos \phi \cos \theta, \\
a'_{22} &= -\sin \varphi \sin \phi + \cos \varphi \cos \phi \cos \theta, \\
a'_{23} &= \cos \phi \sin \theta;
\end{aligned}\right\} \tag{1.3}
$$

$$
\left.\begin{aligned}
a'_{31} &= \sin \varphi \sin \theta, \\
a'_{32} &= \cos \varphi \sin \theta, \\
a'_{33} &= \cos \theta;
\end{aligned}\right\} \tag{1.4}
$$

where the Eulerian angles ϕ, θ, φ are defined by a scheme illustrated on the accompanying Figure IV.1.

A transformation of the inertial to revolving coordinates is similarly governed by the matrix equation

$$
\left\{ \begin{array}{c} x \\ y \\ z \end{array} \right\} = \left\{ \begin{array}{ccc} a''_{11} & a''_{12} & a''_{13} \\ a''_{21} & a''_{22} & a''_{23} \\ a''_{31} & a''_{32} & a''_{33} \end{array} \right\} \left\{ \begin{array}{c} x'' \\ y'' \\ z'' \end{array} \right\}
\tag{1.5}
$$

where the direction cosines

$$
\begin{aligned}
a''_{11} &= \cos u \cos \Omega - \sin u \sin \Omega \cos i, \\
a''_{12} &= -\sin u \cos \Omega - \cos u \sin \Omega \cos i, \\
a''_{13} &= \hphantom{-\sin u \cos \Omega - \cos u} \sin \Omega \sin i;
\end{aligned}
\tag{1.6}
$$

$$
\begin{aligned}
a''_{21} &= \cos u \sin \Omega + \sin u \cos \Omega \cos i, \\
a''_{22} &= -\sin u \sin \Omega + \cos u \cos \Omega \cos i, \\
a''_{23} &= -\sin u \sin \Omega - \cos u \cos \Omega \sin i;
\end{aligned}
\tag{1.7}
$$

$$
\begin{aligned}
a''_{31} &= \hphantom{\cos u} \sin u \sin i, \\
a''_{32} &= \cos u \sin i, \\
a''_{33} &= \hphantom{\cos u \sin} \cos i;
\end{aligned}
\tag{1.8}
$$

where Ω denotes the angle of the nodes (i.e., of intersection of the $z = 0$ and $z'' = 0$ planes measured from the X-axis); i, the inclination of the orbital ($Z'' = 0$) to the invariable ($Z = 0$) plane of the system; and u, the angle between the line of the nodes and the instantaneous position of the radius vector[1].

Accordingly, a transformation from the rotating to the revolving axes obeys the matrix equation

$$
\left\{ \begin{array}{c} x' \\ y' \\ z' \end{array} \right\} = \left\{ \begin{array}{ccc} b''_{11} & b''_{12} & b''_{13} \\ b''_{21} & b''_{22} & b''_{33} \\ b''_{31} & b''_{32} & b''_{23} \end{array} \right\} \left\{ \begin{array}{c} x'' \\ y'' \\ z'' \end{array} \right\},
\tag{1.9}
$$

where the direction cosines b''_{ij} are given by

$$
\left\{ \begin{array}{c} b''_{1j} \\ b''_{3j} \\ b''_{3j} \end{array} \right\} = \left\{ \begin{array}{ccc} a'_{11} & a'_{21} & a'_{31} \\ a'_{12} & a'_{22} & a'_{32} \\ a'_{13} & a'_{23} & a'_{33} \end{array} \right\} \left\{ \begin{array}{c} a''_{1j} \\ a''_{2j} \\ a''_{3j} \end{array} \right\}
\tag{1.10}
$$

for $j = 1, 2, 3$.

Lastly, let the transformation between the doubly- and triply-primed systems of rectangular coordinates be given by the matrix equation

$$
\left\{ \begin{array}{c} x'' \\ y'' \\ z'' \end{array} \right\} = \left\{ \begin{array}{ccc} l_2 & l_1 & l_0 \\ m_2 & m_1 & m_0 \\ n_2 & n_1 & n_0 \end{array} \right\} \left\{ \begin{array}{c} x''' \\ y''' \\ z''' \end{array} \right\},
\tag{1.11}
$$

---

[1] The reader may note that, if, in Equations (1.2)–(1.4) defining the singly-primed direction cosines a'_{ij} we set $\phi = \Omega, \rho = u$, and $\theta = i$, they become identical with the doubly-primed direction cosines a''_{ij}.

where l_0, m_0, n_0 are the direction cosines of the line-of-sight (i.e., the Z'''-axis) in the doubly-primed system of coordinates. Moreover, as the X'''-axis has been defined as a projection of the X''-axis on the $Z''' = 0$ plane, it follows that the direction cosines of the X'''-axis in the doubly-primed system of coordinates on the right-hand side of Equation 1.11) are

$$l_0, \quad l_1 = 0, \quad l_2 = \sqrt{1 - l_0^2} \; ; \tag{1.12}$$

while the remaining direction cosines m_j and n_j ($j = 1, 2$) follow in terms of those for $j = 0$ from the orthogonality conditions

$$\left. \begin{array}{l} l_0 l_1 + m_0 m_1 + n_0 n_1 = 0 , \\ l_0 l_2 + m_0 m_2 + n_0 n_2 = 0 , \\ l_0 l_2 + m_1 m_2 + n_1 n_2 = 0 ; \end{array} \right\} \tag{1.13}$$

which combined with (1.12) yield

$$m_1 = -n_0/l_2, \quad n_1 = m_0/l_2 \tag{1.14}$$

and

$$m_2 = -l_0 m_0/l_2, \quad n_2 = -l_0 n_0/l_2. \tag{1.15}$$

A transformation of the inertial into triply-primed coordinates can be performed with the aid of the matrix equation

$$\left\{ \begin{array}{c} x \\ y \\ z \end{array} \right\} = \left\{ \begin{array}{ccc} a_{11}''' & a_{12}''' & a_{13}''' \\ a_{21}''' & a_{22}''' & a_{23}''' \\ a_{31}''' & a_{32}''' & a_{33}''' \end{array} \right\} \left\{ \begin{array}{c} x''' \\ y''' \\ z''' \end{array} \right\} , \tag{1.16}$$

where

$$\left\{ \begin{array}{c} a_{1j}''' \\ a_{2j}''' \\ a_{3j}''' \end{array} \right\} = \left\{ \begin{array}{ccc} a_{11}'' & a_{12}'' & a_{13}'' \\ a_{21}'' & a_{22}'' & a_{33}'' \\ a_{31}'' & a_{32}'' & a_{33}'' \end{array} \right\} \left\{ \begin{array}{c} l_{3-j} \\ m_{3-j} \\ n_{3-j} \end{array} \right\} \tag{1.17}$$

$j = 1, 2, 3$.

Which ones of the 14 axes specifying our four rectangular systems of coordinates are inertial (i.e., possess directions which are invariant in time)? Only four: namely, X, Y, Z, and Z'''. The singly- and doubly-primed axes rotate and revolve in space—and so do the X'''- and Y'''-axes of the triply-primed system—because the Eulerian angles ϕ, θ, and ϕ contained in the direction cosines a_{ij}' as well as the Keplerian elements Ω, i, and u contained in the a_{ij}'''s are, in general, functions of the time. An exception is the angle of inclination of the invariable ($Z = 0$) plane of the binary system to the plane $Z''' = 0$ tangent to the celestial sphere, which will hereafter be denoted by I. Since both Z- and Z'''-axes are inertial, the angle I should be independent of the time (or its value may change but slowly as our terrestrial observing station and the respective binary may change their relative positions due to their peculiar motions in the Galaxy).

As was already stated before, the inertial XYZ system of coordinates has been fixed so that the XY-plane coincides with the invariable plane of the system; but the directions of the X- and Y-axes in this plane have not been specified so far. In order to remove this arbitrariness, let us constrain the X-axis to lie in the ZZ'''-plane. If so, however, the direction cosines of the Z'''-axis in the inertial frame of reference will be given by

$$a_{13}''' = \sin I, \quad a_{23}''' = 0, \quad a_{33}''' = \cos I; \tag{1.18}$$

and the cosines l_0, m_0, n_0 between the revolving X''-, Y''-, and Z''-axes and the line-of-sight Z''' can be expressed as

$$\begin{aligned}
l_0 &= a_{11}'' \sin I + a_{31}'' \cos I, \\
-m_0 &= a_{12}'' \sin I + a_{32}'' \cos I, \\
n_0 &= a_{13}'' \sin I + a_{33}'' \cos I,
\end{aligned} \right\} \tag{1.19}$$

which on insertion for a_{1j}'' and a_{3j}'' from Equations (1.6) and (1.8) yield

$$l_0 = (\cos u \cos \Omega - \sin u \sin \Omega \cos i) \sin I + \sin u \sin i \cos I, \quad (1.20)$$
$$m_0 = (\sin u \cos \Omega + \cos u \sin \Omega \cos i) \sin I - \cos u \sin i \cos I, \quad (1.21)$$
$$n_0 = \sin \Omega \sin i \sin I + \cos i \cos I. \tag{1.22}$$

If, moreover, we define new angles ψ and j by setting

$$\begin{aligned}
l_0 &= \cos \psi \sin j, \\
m_0 &= \sin \psi \sin j, \\
n_0 &= \cos j,
\end{aligned} \right\} \tag{1.23}$$

the angle j stands evidently for *instantaneous* inclination of the orbital plane to the celestial sphere (obtainable for eclipsing systems from a solution of their light curves for geometrical elements); and ψ, for the "phase angle"—i.e., a true anomaly in a relative elliptic orbit reckoned from the moment of conjunction identified with (say) the primary light minimum.

As is well known, in close binary systems the angles Ω and i in Equations (1.20)–(1.22) will, in general, be also functions of the time—Ω secularly regressing and i oscillating on account of nutation (cf. Kopal, 1989; Chapter VI)—so that, in accordance with Equation (1.22), the actual inclination j of the orbital plane to the celestial sphere should oscillate between $I \pm i$ as Ω runs from 0 to 2π; and will remain secularly constant if $i = 0$ (which can, in turn, be true only if the equators of both components of a close binary system are coplanar with the orbit).

IV.2   Effects Arising from Distortion of the Eclipsing Star

With the geometric preliminaries necessary for the tretment of our problem now completed, let the origin of all systems of coordinates introduced in the preceding section be identified with the centre of mass of one (say, the "primary") component which—in close pairs—may undergo an eclipse by its mate (hereafter to be referred to as the "secondary" component playing the role of the eclipsing star.

Let, moreover, r, θ'', ϕ'' stand for the polar coordinates of an arbitrary point on the surface of the primary component in the revolving system $X''Y''Z''$—such that the rectangular coordinates

$$\left.\begin{array}{rclcl}
x'' & = & r\cos\phi''\sin\theta'' & \equiv & r\lambda, \\
y'' & = & r\sin\phi''\sin\theta'' & \equiv & r\mu, \\
z'' & = & r\cos\theta'' & \equiv & r\nu_j
\end{array}\right\} \tag{2.1}$$

where $r \equiv r(a, \theta'', \phi'')$ stands for an arbitrary radius-vector of the equipotential surface

$$a = \text{constant.} \tag{2.2}$$

Should, moreover, the inclination of the orbital plane to the line of sight be such as to cause the components to eclipse each other at the time of conjunctions, the resulting changes of light will be affected by the distortion of *both* stars of such a couple: a distortion of the component undergoing eclipse will alter both the proportion in size of the apparent segment affected by eclipse, and also the distribution of brightness over the eclipsed part; while a distortion of the secondary (eclipsing) component will cause a corresponding deformation of its shadow cylinder cast by it in the direction of the line of sight. The aim of the present section will be to describe the effects caused by a deformation of this shadow cylinder; while a discussion of the photometric effects going back to a distortion of the primary stars is postponed for the following Section IV.3.

The method by which we propose to do so will be the same as used already in Section III-1. If, in particular, we identify the equatorial plane of the secondary component with that of its orbit, then—to quantities of first order in surficial distortion,

$$r = a\left\{1 - \frac{1}{3}v_2^{(2)}P_2(\nu) + \sum_{j=2}^{4}w_2^{(j)}P_j(\lambda)\right\}, \tag{2.3}$$

where the P_j's stand for the Legendre polynomials of the respective arguments λ and ν as defined by Eqs. (2.1) and were we have abbreviated

$$v_i^{(2)} = \Delta_2\left(1 + \frac{m_{3-i}}{m_i}\right)\left(\frac{a_i}{R}\right)^3 \tag{2.4}$$

and

$$w_i^{(j)} = \Delta_j\frac{m_{3-i}}{m_i}\left(\frac{a_i}{R}\right)^{j+1}, \tag{2.5}$$

$i = 1, 2$, in which $m_{1,2}$ denote the masses of the respective components Δ_j, a coefficient depending only on the internal structure of the respective star; and a_i/R, their fractional dimensions.

Let us return now to the results established in Section III.1 and to the projection of the secondary component on to the plane $Z''' = 0$. For a radius-vector of the form (2.3) its projection ρ on the celestial sphere, defined by Eq. (1.19) of Chapter III—see also Figure III.1—we obtain

$$\rho = a \left\{ 1 - \frac{1}{3} v_2^{(2)} P_2(\nu_2) + \sum_{j=2}^{4} w_2^{(j)} P_j(\lambda_2) \right\}, \tag{2.6}$$

where the arguments λ_2 and ν_2 are defined by the equations

$$\left. \begin{array}{rcl} a_2 \lambda_2 & = & -l_2 a \cos \zeta, \\ a_2 \nu_2 & = & n_1 a \sin \zeta - n_2 a \cos \zeta, \end{array} \right\} \tag{2.7}$$

in which the direction cosines l_2 and $n_{1,2}$ are given by Eqs. (1.12) and (1.14)–(1.15), while the angle ζ continues to be defined by Eq. (1.19), Chapter III (cf. also Figure III.1).

Moreover, over the distorted surface $r \equiv r(a, \theta'', \phi'')$ the element $dx\, dy$ in the Z'''-plane tangent to the celestial sphere can be written as

$$dx\, dy = \rho \rho_a \, da\, d\theta, \tag{2.8}$$

where subscript a on ρ denotes partial differentiation with respect to a. Performing such a differentiation on (2.6)–(2.7), and noting that

$$a \frac{\partial \lambda_2}{\partial a} = \lambda_2 \quad \text{while} \quad a \frac{\partial \nu_2}{\partial a} = \nu_2 \tag{2.9}$$

we find that, to the same degree of accuracy,

$$\begin{aligned} \rho \rho_a = \ a \Big\{ & 1 - \frac{1}{3} v_2^{(2)} [4 P_2(\nu_2) + 1] + \\ & + w_2^{(2)} [4 P_2(\lambda_2) + 1] + \\ & + w_2^{(3)} [5 P_3(\lambda_2) + 3 P_1(\lambda_2)] + \\ & + w_2^{(4)} [6 P_4(\lambda_2) + 5 P_2(\lambda_2) + 1] + \\ & + \cdots \Big\}. \end{aligned} \tag{2.10}$$

Furthermore, as long as the primary component can be regarded as spherical, the Fourier transform $F(u, v)$ defined by Eq.(1.1) of the preceding chapter continues to be given by Eq. (1.15) in which $r = a$. Moreover, the Fourier transform $G(u, v)$ of the "eclipsing" aperture continues to be given by Eq. (1.23) of Chapter III where, however, ρ is now given by (2.6) of this section.

With this in mind, let us proceed to evaluate term-by-term the second integral on the right-hand side of Eq.(1.23), Chapter III, with respect to $d\zeta$. The exponential function behind the integral sign can again be represented by a Jacobi expansion of the form (1.6) of Chapter III; care being merely taken to note that the function ρ as given by (2.6) in the exponential also contains terms arising from distortion of the shadow cylinder. If, therefore, we set $\rho = a(1 + \Delta_2)$, an expansion of $J_j(2\pi q\rho)$ in ascending powers of Δ_2 yields (for $j \geq 0$)

$$
\begin{aligned}
J_j(2\pi q\rho) &= J_j(\xi) + \left\{ \xi \frac{d}{d\xi} J_j(\xi) \right\} \Delta_2 + \cdots \\
&= J_j(\xi) + \{ j J_j(\xi) - \xi J_{j+1}(\xi) \} \Delta_2 + \cdots ,
\end{aligned}
\tag{2.11}
$$

by the recursion formula (3.1) of Chapter III, in which (to the first order in small quantities)

$$
\Delta_2 \equiv -\frac{1}{3} v_2^{(2)} P_2(\nu_2) + \sum_{j=2}^{4} w_2^{(j)} P_j(\lambda_2),
\tag{2.12}
$$

and where we have set

$$
\xi = 2\pi qa .
\tag{2.13}
$$

Accordingly,

$$
\begin{aligned}
\int_{-\pi}^{\pi} e^{-2\pi iq\rho \cos(\zeta - \phi)} \rho \rho_a \, d\zeta &= \pi \left\{ [\xi J_1(\xi) - 2J_0(\xi)] - \right. \\
&\quad - \frac{3}{2} \left(\frac{al_2}{a_2} \right)^2 [\xi J_1(\xi) - 4J_0(\xi)] + \\
&\quad \left. + \frac{3}{2} \left(\frac{al_2}{a_2} \right)^2 [\xi J_3(\xi) - 6J_2(\xi)] \cos 2\phi \right\} w_2^{(2)} + \\[2mm]
&\quad + i\pi \left\{ 3 \left[\frac{al_2}{a_2} \{ \xi J_2(\xi) - 4J_1(\xi) \} - \right. \right. \\
&\quad \left. - \frac{5}{4} \left(\frac{al_2}{a_2} \right)^3 \{ \xi J_2(\xi) - 6J_1(\xi) \} \right] \cos \phi + \\
&\quad \left. + \frac{5}{4} \left(\frac{al_2}{a_2} \right)^3 [\xi J_4(\xi) - 8J_3(\xi)] \cos 3\phi \right\} w_2^{(3)} - \\[2mm]
&\quad - \frac{\pi}{4} \left\{ 3[\xi J_1(\xi) - 2J_0(\xi)] - 15 \left(\frac{al_2}{a_2} \right)^2 [\xi J_1(\xi) - 4J_0(\xi)] + \right. \\
&\quad + \frac{105}{8} \left(\frac{al_2}{a_2} \right)^4 [\xi J_1(\xi) - 6J_0(\xi)] + \\
&\quad + 15 \left(\frac{al_2}{a_2} \right)^2 [\xi J_3(\xi) - 6J_2(\xi)] \cos 2\phi - \\
&\quad - \frac{35}{2} \left(\frac{al_2}{a_2} \right)^4 [\xi J_3(\xi) - 8J_2(\xi)] \cos 2\phi + \\
&\quad \left. + \frac{35}{8} \left(\frac{al_2}{a_2} \right)^4 [\xi J_5(\xi) - 10J_4(\xi)] \cos 4\phi \right\} w_2^{(4)} -
\end{aligned}
$$

$$-\frac{\pi}{3}\left\{[\xi J_1(\xi) - 2J_0(\xi)]-\right.$$

$$-\frac{9}{2}\left(\frac{a}{a_2}\right)^2(1 - n_0^2)[\xi J_1(\xi) - 4J_0(\xi)] + \tag{2.14}$$

$$+\frac{9}{2}\left(\frac{a}{a_2}\right)^2(1 - n_0^2)[\xi J_3(\xi) - 6J_2(\xi)]\cos 2\phi\left.\right\}v_2^{(2)}$$

where the angle ϕ continues to be defined by Eq. (1.5) of Chapter III, and i stands for the imaginary unit.

Two integrations remain yet to be performed before we obtain the convolution integral (1.26) of Chapter III. The one with respect to $d\phi$ follows readily from

$$\int_{-\pi}^{\pi} e^{-2\pi i \delta u}\cos m\phi\, d\phi \equiv \int_{-\pi}^{\pi} e^{-2\pi i \delta q\,\cos\phi}\cos m\phi =$$

$$= 2\pi(-i)^m\, J_m(2\pi\delta q), \tag{2.15}$$

of which Eqs.(1.29) and (5.9)of Chapter III represent particular cases for $m = 0$ and 1. Moreover, in order to carry out the integration with respect to a, all we need to do is to generalize Eq. (1.25) of Chapter III by setting

$$a = a_2 y, \tag{2.16}$$

which renders

$$\int_0^{a_2} j_j(\xi)a^{j+1}da = a_2^{j+1}\int_0^1 J_j(xy)y^{j+1}dy =$$

$$= \left(\frac{a_2^{j+1}}{x}\right)J_{j+1}(x), \tag{2.17}$$

in which we have abbreviated

$$x \equiv 2\pi q a_2, \tag{2.18}$$

in agreement with Eq. (1.33) of Chapter III.

If, lastly, we insert for $F(u, v)$ in (1.30) from (1.15) of Chapter III and set

$$(2\pi a_2)^2 q\, dq = x\, dx \tag{2.19}$$

in agreement with (2.18), the photometric effects f_2 of distortion of the shadow cylinder cast by the secondary component can be expressed as

$$f_2 \equiv \sum_{n=0}^{N} C^{(n)}\, f_2^{(n)}, \tag{2.20}$$

where the $C^{(n)}$s' stand for the normalized limb-darkened coefficients defined by Eqs. (1.11)–(1.12) of Chapter II, and

$$k^{n+2}f_2^{(n)} = -\{n_1^2 I_{1,n}^0 + n_2^2 I_{-1,n}^2 - \frac{1}{3}I_{-1,n}^0\}v_2^{(2)} +$$

$$+ \{3l_2^2 I_{-1,n}^2 - I_{-1,n}^0\} w_2^{(2)} +$$
$$+ \{5l_2^3 I_{-1,n}^3 - 3l_2 I_{-1,n}^1\} w_2^{(3)} +$$
$$+ \frac{1}{4}\{35l_2^4 I_{-1,n}^4 - 30l_2^2 I_{-1,n}^2 + 3I_{-1,n}^0\} w_2^{(4)} +$$
$$+ \cdots , \qquad (2.21)$$

where $k \equiv a_1/a_2$ stands for the ratio of the radii of the primary and secondary components, $v_2^{(2)}$ and $w_2^{(j)}$ are the coefficients of distortion of the shadow-casting secondary as defined by Eqs.(2.4)–(2.5); while l_2 and $n_{1,2}$ are the direction cosines as given by Eqs. (1.12)–(1.15).

The eclipse functions arising from tidal distortion are given by

$$I_{-1,n}^0 = 2^{\nu-1}\Gamma(\nu)\int_0^\infty \frac{J_\nu(kx)}{(kx)^\nu}J_0(x)J_0(hx)x\,dx, \qquad (2.22)$$

$$I_{-1,n}^1 = 2^{\nu-1}\Gamma(\nu)\int_0^\infty \frac{J_\nu(kx)}{(kx)^\nu}J_1(x)J_1(hx)x\,dx, \qquad (2.23)$$

$$I_{-1,n}^2 = 2^{\nu-2}\Gamma(\nu)\int_0^\infty \frac{J_\nu(kx)}{(kx)^\nu}\{J_0(x)J_0(hx) + $$
$$+ J_2(x)J_2(hx)\}\,x\,dx, \qquad (2.24)$$

$$I_{-1,n}^3 = 2^{\nu-3}\Gamma(\nu)\int_0^\infty \frac{J_\nu(kx)}{(kx)^\nu}\{3J_1(x)J_1(hx)+$$
$$+ J_3(x)J_3(hx)\}\,x\,dx , \qquad (2.25)$$

$$I_{-1,n}^4 = 2^{\nu-4}\Gamma(\nu)\int_0^\infty \frac{J_\nu(kx)}{(kx)^\nu}\{3J_0(x)J_0(hx)+$$
$$+ 4J_2(x)J_2(hx) + J_4(x)J_4(hx)\}\,x\,dx; \qquad (2.26)$$

where the parameters h and k continue to be given by Eqs. (1.35) of Chapter III and, in accordance with Eq. (1.16) of that chapter,

$$\nu = \frac{n+2}{2}; \qquad (2.27)$$

while the only new term arising from the rotational distortion can be reduced to

$$2\delta\nu\, I_{1,n}^0 = r_2 I_{-1,n+2}^1 \qquad (2.28)$$

or

$$2\nu\, J_{1,n}^0 = J_{-1,n+2}^1 \qquad (2.29)$$

by (2.23) and (2.24) of Chapter II.

The integrals on the right-hand sides of Eqs. (2.22)–(2.26) differ from those defining the associated alpha-functions α_n^0 by the fact that the orders of two

Bessel functions in each expression are identical (say, m). If so, then (cf. e.g., Erdélyi et al, 1954; p.52, No.34)

$$\int_0^\infty \frac{J_\nu(kx)}{(kx)^\nu} J_m(x) J_m(hx) x\, dx =$$
$$= h^{\nu-1} k^{-2\nu} (1-\mu^2)^{(n+1)/4} P_{m-\frac{1}{2}}^{-(n+1)/2}(\mu), \qquad (2.30)$$

where the argument μ of the Legendre functions $P_{m-\frac{1}{2}}^{-(n+1)/2}(\mu)$ is given by

$$\mu \equiv \frac{a_2^2 - a_1^2 + \delta^2}{2\delta a_2} = 1 - 2\kappa^2 \qquad (2.31)$$

in agreement with Eq. (2.55) of Chapter II; and, accordingly,

$$\sqrt{2\pi}\, I_{-1,n}^0 = \Gamma(\nu)(2h)^{\nu-1}(\sin v)^{\nu-\frac{1}{2}} P_{-\frac{1}{2}}^{\frac{1}{2}-\nu}(\cos v), \qquad (2.32)$$

$$\sqrt{2\pi}\, I_{-1,n}^1 = \Gamma(\nu)(2h)^{\nu-1}(\sin v)^{\nu-\frac{1}{2}} P_{\frac{1}{2}}^{\frac{1}{2}-\nu}(\cos v), \qquad (2.33)$$

$$2\sqrt{2\pi}\, I_{-1,n}^2 = \Gamma(\nu)(2h)^{\nu-1}(\sin v)^{\nu-\frac{1}{2}} \left\{ P_{\frac{3}{2}}^{\frac{1}{2}-\nu}(\cos v) + P_{-\frac{1}{2}}^{\frac{1}{2}-\nu}(\cos v) \right\} \quad (2.34)$$

$$4\sqrt{2\pi}\, I_{-1,n}^3 = \Gamma(\nu)(2h)^{\nu-1}(\sin v)^{\nu-\frac{1}{2}} \left\{ P_{\frac{5}{2}}^{\frac{1}{2}-\nu}(\cos v) + 3P_{\frac{1}{2}}^{\frac{1}{2}-\nu}(\cos v) \right\} \quad (2.35)$$

$$8\sqrt{2\pi}\, I_{-1,n}^4 = \Gamma(\nu)(2h)^{\nu-1}(\sin v)^{\nu-\frac{1}{2}} \times$$
$$\times \left\{ P_{\frac{7}{2}}^{\frac{1}{2}-\nu}(\cos v) + 4P_{\frac{3}{2}}^{\frac{1}{2}-\nu}(\cos v) + 3P_{-\frac{1}{2}}^{\frac{1}{2}-\nu}(\cos v) \right\}, \quad (2.36)$$

in which we have abbreviated $\cos v = \mu$.

Lastly, since the Legendre functions involved are known to satisfy the recursion relation

$$P_{\frac{3}{2}}^{\frac{1}{2}-\nu}(\cos v) = P_{-\frac{1}{2}}^{\frac{1}{2}-\nu}(\cos v) - 2(\sin v) P_{\frac{1}{2}}^{-\frac{1}{2}-\nu}(\cos v), \qquad (2.37)$$

the rotational term $I_{1,n}^0$—related with $I_{-1,n+2}^1$ by means of Eq. (2.28)—can with the aid of (2.33) and (2.37) be rewritten as

$$\sqrt{2\pi}\, I_{1,n}^0 = \Gamma(\nu)(2h)^{\nu-1}(\sin v)^{\nu+\frac{1}{2}} P_{\frac{1}{2}}^{-\frac{1}{2}-\nu}(\cos v). \qquad (2.38)$$

The reader may note that the foregoing Equations (2.32) and (2.33) are already known to us from Eqs. (2.42)–(2.43) and (2.52)–(2.53) established in Chapter II; but those for $I_{-1,n}^m$ with $m > 1$ are new.

IV.3 Effects Arising from Distortion of the Eclipsed Star

In turning to investigate the photometric effects of distortion of the component undergoing eclipse, we find ourselves facing a more complicated situation; for whereas in the preceding section we were concerned only with the distortion of the shadow cylinder cast by the secondary (eclipsing) component on to the celestial sphere while the primary (eclipsed) star was regarded as a sphere, the effects arising from the primary's distortion will include not only more complicated geometry of the visible segments, but also those affecting the distribution of brightness over this segment caused by limb- and gravity-darkening.

In more specific terms, let the cosine of the angle of foreshortening γ introduced in Equation (1.2) of Chapter II be expressed as

$$\cos \gamma = l l_0 + m m_0 + n n_0 , \tag{3.1}$$

where l_0, m_0, n_0 stand for the direction cosines of the line of sight as defined by Eqs. (1.23) in Section IV.1 and l, m, n are the direction cosines of a normal to an equipotential surface of the form (2.2). The latter are, in turn, given by the equation

$$l, m, n = \frac{a_{x'',y'',z''}}{\sqrt{a_{x''}^2 + a_{y''}^2 + a_{z''}^2}} , \tag{3.2}$$

which on evaluation (cf. sec. V.1 of Kopal, 1989) yield, correctly to quantities of first order in surficial distortion,

$$\cos \gamma = L \left\{ 1 + \frac{1}{3} v_1^{(2)} \left[\frac{n_0}{L} - \nu_1 \right] P_2'(\nu_1) - \left[\frac{l_0}{L} - \lambda_1 \right] \sum_{j=2}^{4} w_1^{(j)} P_j'(\lambda_1) \right\} , \tag{3.3}$$

where (in agreement with Eq. (1.13) of Chapter III)

$$L = \lambda'' l_0 + \mu'' m_0 + \nu'' n_0 = \sqrt{1 - \left(\frac{a}{a_1} \right)^2} , \tag{3.4}$$

in which a_1 specifies the position of the primary's surface[2]; and the coefficients $v_1^{(2)}$, $w_1^{(j)}$ of its distortion continue to be given by Eqs. (2.4) and (2.5) of this chapter for $i = 1$; while the arguments λ_1 and ν_1 in the plane $Z'' = 0$ of the celestial sphere are given by the equations

$$a_1 \lambda_1 = l_0 \sqrt{a_1^2 - a^2} \qquad\qquad + l_2 a \cos \theta$$
$$a_1 \nu_1 = n_0 \sqrt{a_1^2 - a^2} \quad + n_1 \sin \theta \quad + n_2 a \cos \theta, \tag{3.5}$$

---

[2] L signifies the direction cosine denoted on p.224 of Kopal (1989) by N.

in which the angle θ continues to be defined by Eq. (1.3) of Chapter III (see also Fig. III.1)—replacing Eqs. (2.7) of the preceding section—and, lastly, the primes on the Legendre polynomials P_j on the r.h.s. of Eq. (3.3) denote differentiation with respect to their arguments.

Moreover, the radiative transfer, in distorted stars will cause the flux of light H emerging normally to equipotential surfaces (cf., e.g., Sec. VII.2 of Kopal, 1989) to vary as

$$H = H_0 \left\{ 1 + \tau \left(\frac{1}{r_a} - 1 \right) + \cdots \right\} =$$
$$= \left\{ 1 + \frac{g - g_0}{g_0} \tau + \cdots \right\}, \tag{3.6}$$

where the variation g over the distorted surface should vary as

$$\frac{g - g_0}{g_0} = \frac{1}{3} \left\{ \frac{5}{\Delta_2} - 1 \right\} v_1^{(2)} P_2(\nu_1) - \sum_{j=2}^{4} \left\{ \frac{2j + 1}{\Delta_j} - j + 1 \right\} w_2^{(j)} P_j(\lambda_1), \tag{3.7}$$

where the coefficient τ of gravity-darkening for black-body radiation will (in accordance with Planck's Law) be given as a function of the temperature T and wavelength λ by

$$\tau \equiv \frac{c_2/\lambda T}{4[1 - \exp(-c_2/\lambda T)]}, \tag{3.8}$$

where $c_2 = 1.438$ cm. deg; and Δ_j (as before) a coefficient depending only on the internal structure of the respective star, the reader will find extensively discussed in Chapter IV of Kopal (1989).

If so, the Fourier transform $F(u, v)$ as given by Equation (1.1) of Chapter III, can be rewritten as

$$F(u, v) = \sum_{n=0}^{N} C^{(n)} F^{(n)}(q, \phi), \tag{3.9}$$

where the coefficients $C^{(n)}$ of limb-darkening continue to be given by Eqs. (1.11)–(1.12), Chapter II, and

$$F^{(n)}(q, \phi) \equiv \int_0^{a_1} \int_0^{\pi} \cos^n \gamma \left(\frac{H}{H_0} \right) \times$$
$$\times e^{-2\pi i q \rho \cos(\theta - \phi)} \rho \rho_a \, da \, d\theta, \tag{3.10}$$

in which $\cos \gamma$ and H/H_0 continue to be given by Eqs. (3.3) and (3.6), and ρ stands for the projection of an arbitrary radius-vector r of the primary's surface on the celestial sphere, as defined by Eq. (1.29) of Chapter III, and given by Eq. (2.6) in which the indices i of $v_i^{(2)}$, $w_i^{(j)}$ and λ_i, ν_i have now been identified with $i = 1$.

In order to evaluate the integrals on the right-hand side of the preceding Equation (3.10), let us expand again the exponential behind the integral sign in the Jacobi series (1.6) of Chapter III, and express the argument $2\pi q\rho$ of the Bessel functions involved in (3.6) in terms of $2\pi qa \equiv \xi$ by means of Eq. (2.11) of the present chapter: doing so we find that, correctly to terms of first order in $v_1^{(2)}$ and $w_1^{(j)}$,

$$\cos^n \gamma (H/H_0) \exp\{-2\pi i q\rho \cos(\theta - \phi)\}\rho\rho_a =$$

$$= aL^n \{J_0(\xi) + \exp\{-i\xi \cos(\theta - \phi)\} \times$$

$$\times \left[-\frac{1}{3}v_1^{(2)} \left\{ \left[2(n+2) - \tau \left(\frac{5}{\Delta_2} - 1 \right) \right] P_2(\nu_1) + (n+1) \left[1 - 3\frac{n_0}{L} P_1(\nu_1) \right] \right\} +$$

$$+ \sum_{j=2}^{4} w_1^{(j)} \left\{ \left[2 - \tau \left(\frac{2j+1}{\Delta_j} - j + 1 \right) \right] P_j(\lambda_1) +$$

$$+ (n+1) \left[jP_j(\lambda_1) + P_{j-1}'(\lambda_1) - \frac{l_0}{L} P_j'(\lambda_1) \right] \right\} \right] +$$

$$+ aL^n \left\{ -\xi J_1(\xi) + 2 \sum_{m=1}^{\infty} [2i J_{2m}(\xi) - \xi J_{2m+1}(\xi)] \cos 2m \left(\frac{\pi}{2} + \theta - \phi \right) -$$

$$- 2i \sum_{m=0}^{\infty} [(2m+1)J_{2m+1}(\xi) - \xi J_{2m+2}(\xi)] \sin(2m+1) \left(\frac{\pi}{2} + \theta - \phi \right) \right\} \times$$

$$\times \left\{ -\frac{1}{3}v_1^{(2)} P_2(\nu_1) + \sum_{j=2}^{4} w_1^{(j)} P_j(\lambda_1) \right\}, \tag{3.11}$$

where λ_1, ν_1 are given by Eqs. (3.5).

By the addition theorem for Legendre polynomials

$$P_1(\nu_1) = n_0 P_1(L) + (n_1 \sin \theta + n_2 \cos \theta) P_1^1(L), \tag{3.12}$$

$$P_2(\nu_1) = P_2(n_0) P_2(L) - n_0 (n_1 \sin \theta + n_2 \cos \theta) P_2^1(L) +$$

$$+ \frac{1}{2}[2n_1 n_2 \sin 2\theta + (n_2^2 - n_1^2) \cos 2\theta] P_2^2(L); \tag{3.13}$$

while

$$P_j(\lambda_1) = P_j(l_0) P_j(L) + 2 \sum_{m=1}^{j} \frac{(j-m)!}{(j+m)!} P_j^m(l_0) P_j^m(L) \cos m\theta. \tag{3.14}$$

In consequence, an integration with respect to θ between $\pm\pi$ yields

$$\int_{-\pi}^{\pi} e^{-i\xi \cos(\theta - \phi)} d\theta = 2\pi J_0(\xi), \tag{3.15}$$

$$\int_{-\pi}^{\pi} e^{-i\xi \cos(\theta - \phi)} P_1(\lambda_1) d\theta = 2\pi \{P_1(l_0) P_1(L) J_0(\xi) -$$

$$- i P_1^1(l_0) P_1^1(L) J_1(\xi) \cos \phi\}, \tag{3.16}$$

$$\int_{-\pi}^{\pi} e^{-i\xi \cos(\theta-\phi)} P_2(\lambda_1) d\theta = 2\pi \{P_2(l_0)P_2(L)J_0(\xi)-$$

$$-\frac{1}{3}i\, P_2^1(l_0)P_2^1(L)J_1(\xi) \cos \phi +$$

$$+\frac{1}{12}P_2^2(l_0)P_2^2(L)J_2(\xi) \cos 2\phi\}, \qquad (3.17)$$

$$\int_{-\pi}^{\pi} e^{-i\xi \cos(\theta-\phi)} P_3(\lambda_1) d\theta = 2\pi \{P_3(l_0)P_3(L)J_0(\xi)-$$

$$-\frac{i}{6}P_3^1(l_0)P_3^1(L)J_1(\xi) \cos \phi +$$

$$+\frac{1}{60}P_3^2(l_0)P_3^2(L)J_2(\xi) \cos \phi -$$

$$-\frac{i}{360}P_3^3(l_0)P_3^3(L)J_3(\xi) \cos 3\phi\}, \qquad (3.18)$$

$$\int_{-\pi}^{\pi} e^{-i\xi \cos(\theta-\phi)} P_4(\lambda_1) d\theta = 2\pi \{P_4(l_0)P_4(L)J_0(\xi)-$$

$$-\frac{i}{10}P_4^1(l_0)P_4^1(L)J_1(\xi) \cos \phi +$$

$$+\frac{1}{180}P_4^2(l_0)P_4^2(L)J_2(\xi) \cos 2\phi -$$

$$-\frac{i}{2520}P_4^3(l_0)P_4^3(L)J_3(\xi) \cos 3\phi +$$

$$+\frac{1}{20160}P_4^4(l_0)P_4^4(L)J_4(\xi) \cos 4\phi\} \qquad (3.19)$$

and

$$\int_{-\pi}^{\pi} e^{-i\xi \cos(\theta-\phi)} P_1(\nu_1) d\theta = 2\pi \{P_1(n_0)P_1(L)J_0(\xi)+$$

$$+ i(n_1 \sin \phi + n_2 \cos \phi)P_1^1(L)J_1(\xi)\}, \qquad (3.20)$$

$$\int_{-\pi}^{\pi} e^{-i\xi \cos(\theta-\phi)} P_2(\nu_1) d\theta = 2\pi \{P_2(n_0)P_2(L)J_0(\xi)+$$

$$+ in_0(n_1 \sin \phi + n_2 \cos \phi)P_2^1(L)J_1(\xi) + \qquad (3.21)$$

$$+\frac{1}{4}[(n_1^2 - n_2^2) \cos \phi - 2n_1 n_2 \sin \phi]P_2^2(L)J_2(\xi)\};$$

while integrals of the derivatives $P_j'(\lambda)$ with respect to λ can be deduced by a combination of the foregoing expressions (3.15)–(3.19) from the identity

$$\lambda_1 P_j'(\lambda_1) = j\, P_j(\lambda_1) = P_{j-1}'(\lambda_1) . \qquad (3.22)$$

The remaining integrals with respect to θ, arising from the last term on the right-hand side of Equation (3.11) are purely trigonometric and offer no difficulty:

in fact, from Eqs. (3.17)–(3.21) it follows readily that, for $m \geq 0$,

$$\int_{-\pi}^{\pi} P_j(\lambda_1) \cos 2m \left(\frac{\pi}{2} + \theta - \phi \right) d\theta =$$

$$= 2\pi(-1)^m \frac{(j - 2m)!}{(j + 2m)!} P_j^{2m}(l_0) P_j^{2m}(L) \cos 2m\phi \qquad (3.23)$$

and

$$\int_{-\pi}^{\pi} P_j(\lambda_1) \sin(2m + 1) \left(\frac{\pi}{2} + \theta - \phi \right) d\theta =$$

$$= 2\pi(-1)^m \frac{(j - 2m - 1)!}{(j + 2m + 1)!} P_j^{2m+1}(l_0) P_j^{2m+1}(L) \cos(2m + 1)\phi; \qquad (3.24)$$

(it being understood that $P_j^i \equiv 0$ for $i > j$); while

$$\int_{-\pi}^{\pi} P_2(\nu_1) d\theta = 2\pi P_2(n_0) P_2(L), \qquad (3.25)$$

$$\int_{-\pi}^{\pi} P_2(\nu_1) \sin \left(\frac{\pi}{2} + \theta - \phi \right) d\theta =$$

$$= -\pi n_0 (n_1 \sin \phi + n_2 \cos \phi) P_2^1(L) \qquad (3.26)$$

and

$$\int_{-\pi}^{\pi} P_2(\nu_1) \cos 2 \left(\frac{\pi}{2} + \theta - \phi \right) d\theta = \qquad (3.27)$$

$$= \frac{\pi}{4} \{ (n_1^2 - n_2^2) \cos 2\phi - 2n_1 n_2 \sin 2\phi \} P_2^2(L).$$

Having thus obtained all the requisite integrals of Eq. (3.11) with respect to θ, the only task still outstanding is to integrate the expression (3.10) for $F^{(n)}(q, \phi)$ with respect to a. This too can be accomplished in the same way as in the preceding section: namely, by noting that if we set

$$a = a_1 y, \qquad (3.28)$$

the expression

$$\int_0^1 (1 - y^2)^{\nu-1} J_j(xy) y^{j+1} dy = 2^{\nu-1} \Gamma(\nu) x^{-\nu} J_{j+\nu}(x) \qquad (3.29)$$

obtains as a generalization of Eq. (2.17), where $x \equiv 2\pi q r_2$ continues to be given by Eq. (2.18).

Having done so, we have thereby established the explicit form of the Fourier transform $F(u, v)$ of a distorted source undergoing eclipse as functions of q and

ϕ (as defined by Eq. (1.3) of Chapter III). To evaluate the corresponding loss of light f_1 arising from the distortion of the primary component, defined as

$$f_1 = \sum_{n=0}^{N} C^{(n)} f_1^{(n)}, \tag{3.30}$$

it remains yet for us to evaluate the convolution integral of the functions $F(u, v)$ and $G(u, v)$ as defined by Equation (1.26) of Chapter III. Correctly to terms of first order in surficial distortion of the star undergoing eclipse—i.e., if we agree to neglect the cross-products of the coefficients $v_i^{(2)}$ and $w_i^{(j)}$ which specify distortion of both components as well as their squares—the Fourier transform $G(u, v)$ of the shadow cone cast by the eclipsing star will, however, retain the same form as established already in Section III.1 and given by Equation (1.25) of that chapter.

In order to establish the respective convolution, the product $F(u, v)G(u, v)$ has still to be integrated over the entire plane—i.e., within $\pm\pi$ in ϕ, and $(0, \infty)$ in q (or x). The former task can again be entirely accomplished by a resort to Equation (2.15) used already in the preceding section of this chapter. As regards the only remaining integration with respect to q, we note that the expressions so obtained turn out to be identical with those defining certain families of special functions we met already in Chapter III at earlier stages of our discussion, and which will re-appear to enable us to express our present results in more concise form: namely, if we take advantage of numerous identities obtaining among the associated Legendre polynomials P_n^m, we find that

$$\begin{aligned}
f_1^{(n)} &= a_2^{(0)} \left\{ \frac{1}{3} v_1^{(2)} P_2(n_0) - w_1^{(2)} P_2(l_0) \right\} + \\
&\quad + a_2^{(1)} \left\{ \frac{1}{3} v_1^{(2)} n_0 n_2 + w_1^{(2)} P_2^1(l_0) \right\} + \\
&\quad + a_2^{(2)} \left\{ \frac{1}{3} v_1^{(2)} (n_1^2 - n_2^2) - w_1^{(2)} P_2^2(l_0) \right\} + \\
&\quad + \sum_{m=0}^{3} w_1^{(3)} a_3^{(m)} P_3^m(l_0) + \\
&\quad - \sum_{m=0}^{4} w_1^{(4)} a_4^{(m)} P_4^m(l_0), \tag{3.31}
\end{aligned}$$

where

$$a_2^{(0)} = \frac{1}{2}\{3(\beta_2 - n)\alpha_{n+2}^0 - (\beta_2 - 2n)\alpha_n^0 + n\alpha_{n-2}^0\}, \tag{3.32}$$

$$a_2^{(1)} = \{(b_2 - n)\alpha_n^1\}, \tag{3.33}$$

$$a_2^{(2)} = \frac{1}{2}\left\{ \frac{\beta_2 - 3h - 5}{h+1} \left[\frac{a_1}{\delta}\alpha_{n+2}^1 - \Im_{-1,n+2}^0 \right] + \right. \\
\left. + \left[\frac{a_1}{\delta}\alpha_n^1 - \Im_{-1,n}^0 \right] \right\}; \tag{3.34}$$

$$a_3^{(0)} = \frac{1}{4}\left\{[6\beta_3 - 5(5h - 2)]\alpha_{n+1}^0 - 5(2\beta_3 - 5h)\alpha_{n+3}^0\right\}, \tag{3.35}$$

$$a_3^{(1)} = \frac{1}{8}\left\{5(\beta_3 - 4h - 6)\alpha_{n+2}^1 + (\beta_3 + 11h + 7)\alpha_n^1 - n\alpha_{n-2}^1\right\}, \tag{3.36}$$

$$a_3^{(2)} = \frac{1}{8}\left\{\frac{\beta_3 - 4h - 6}{h + 2}\left[\frac{a_1}{\delta}\alpha_{n+3}^1 - \Im_{-1,n+3}^0\right] + \right.$$
$$\left. +2\left[\frac{a_1}{\delta}\alpha_{n+1}^1 - \Im_{-1,n+1}^0\right]\right\}, \tag{3.37}$$

$$a_3^{(3)} = \frac{1}{24}\left\{\frac{\beta_3 - 4h - 6}{h + 1}\left[\frac{2}{h + 3}\frac{a_1}{\delta}\left(\frac{a_1}{\delta}\alpha_{n+4}^1 - \Im_{-1,n+4}^0\right) - \right.\right.$$
$$\left. -(\nu + 2)\alpha_{n+2}^1 + \nu\alpha_n^1\right] +$$
$$\left. +\frac{2}{h + 1}\frac{a_1}{\delta}\left(\frac{a_1}{\delta}\alpha_{n+2}^1 - \Im_{-1,n+2}^0\right) - (\nu + 1)\alpha_n^1 + \frac{n}{2}\alpha_{n-2}^1\right\}; \tag{3.38}$$

$$a_4^{(0)} = \frac{1}{8}\left\{35(\beta_4 - 3n)\alpha_{n+4}^0 - 15(2\beta_4 - 9n)\alpha_{n+2}^0 + \right.$$
$$\left. +3(\beta_4 - 9n)\alpha_n^0 - 3n\alpha_{n-2}^0\right\}, \tag{3.39}$$

$$a_4^{(1)} = \frac{1}{4}\left\{(\beta_4 - 4h - 3)(7\alpha_{n+3}^1 - 3\alpha_{n+1}^1) + 3h(5\alpha_{n+1}^1 - \alpha_{n-1}^1)\right\}, \tag{3.40}$$

$$a_4^{(2)} = -\frac{1}{24}\left\{\frac{7(\beta_4 - 9\nu)}{\nu + 1}\left[\frac{a_1}{\delta}\alpha_{n+4}^1 - \Im_{-1,n+4}^0\right] - \right.$$
$$-\frac{\beta_4 - 8h - 7}{\nu}\left[\frac{a_1}{\delta}\alpha_{n+2}^1 - \Im_{-1,n+2}^0\right] -$$
$$\left. -\left[\frac{a_1}{\delta}\alpha_n^1 - \Im_{-1,n}^0\right]\right\}, \tag{3.41}$$

$$a_4^{(3)} = -\frac{1}{48}\left\{\frac{\beta_4 - 5h - 7}{(h + 2)(h + 4)}\left[\frac{8a_1}{\delta}\left(\frac{a_1}{\delta}\alpha_{n+5}^1 - \Im_{-1,n+5}^0\right) - \right.\right.$$
$$\left. -(h + 4)\left((h + 6)\alpha_{n+3}^1 - (h + 2)\alpha_{n+1}^1\right)\right] +$$
$$+\frac{2}{h + 2}\left[\frac{8a_1}{\delta}\left(\frac{a_1}{\delta}\alpha_{n+3}^1 - \Im_{-1,n+3}^0\right) - \right.$$
$$\left. -(h + 2)\left(9h + 4)\alpha_{n+1}^1 - h\alpha_{n-1}^1\right)\right]\right\}, \tag{3.42}$$

$$a_4^{(4)} = \frac{1}{192(h + 1)(h + 3)(h + 5)}\left\{24\frac{a_1^2}{\delta^2}[(\beta_4 - 5h - 7)\times\right.$$
$$\times\left(\frac{a_1}{\delta}\alpha_{n+6}^1 - \Im_{-1,n+6}^0\right) - (h + 5)\left(\frac{a_1}{\delta}\alpha_{n+4}^1 - \Im_{-1,n+4}^0\right)\right] -$$

$$- (h + 7)(\beta_4 - 5h - 7) \left(\frac{4a_1}{\delta} \alpha_{n+4}^1 - \mathfrak{S}_{-1,n+4}^0 \right) +$$

$$+ (h + 3) \left[(\beta_4 - 4h - 2) \left(\frac{4a_1}{\delta} \alpha_{n+2}^1 - \mathfrak{S}_{-1,n+2}^0 \right) - \right.$$

$$\left. - (h + 1) \left(\frac{4a_1}{\delta} \alpha_n^1 - \mathfrak{S}_{-1,n} \right) \right] \Big\} , \tag{3.43}$$

where it may be noted that

$$\frac{a_1}{\delta} \alpha_n^1 = 2 \left(\frac{a_2}{a_1} \right)^{n+2} I_{1,n}^0 \tag{3.44}$$

by Eq. (2.23) of Chapter II; and

$$\beta_j = \left\{ \frac{2j + 1}{\Delta_j} + 1 - j \right\} \tau \simeq (j + 2)\tau, \quad j = 2, 3, 4, \tag{3.45}$$

for centrally-condensed configurations ($\Delta_j \simeq 1$); and $h \equiv n + 1$.

The foregoing Equations (3.32)–(3.44) hold good, as they stand, for any non-negative values of n including zero—if note is taken of the fact that, for $n = 0$,

$$\lim_{n \to 0} n\alpha_{n-2}^0 = \frac{2}{\pi} \cos^{-1} \frac{s}{r_1} \tag{3.46}$$

and

$$\lim_{n \to 0} n\alpha_{n-2}^1 = \frac{2}{\pi} \sqrt{1 - \frac{s^2}{r_1^2}} , \tag{3.47}$$

where s continues to be defined by Eq. (1.7) of Chapter II (for a proof of these limits cf. p. 89 of Kopal, 1979).

A comparison of our present equation (3.25) with Eqs. (2.31) + (2.32), Chapter II of Kopal (1979); or with (3.44) + (3.45), Chapter VII of Kopal (1989) demonstrates clearly the superiority of the approach to our problem initiated in this chapter. For whereas our former approach called for the explicit use of the special functions of eclipses of the types α_n^m and $\mathfrak{S}_{-1,n}^m$ for $m = 0(1)4$ to describe first-order effects of surficial distortion of the primary (eclipsed) component of close binary systems, *our present approach restricts this requirement to the α_n^m's for only $m = 0$ and 1* (since the $\mathfrak{S}_{-1,n}^0$'s are expressible in terms of the α_n^0's by means of recursion formula (2.17) of Chapter II).

Moreover—as was shown already in Section III.1—the α_n^m's for $m = 0$ and 1 are also the easiest ones to evaluate in terms of the Hankel functions for the elements of the eclipse; so that the numerical work (by hand, or machine) entailed in applications to practical cases to be outlined in the next chapter becomes much less laborious than was the case before—the less so, the larger the value of m (as only an increasingly smaller fraction of special functions employed in our previous treatment of the subject (cf. Kopal, 1979) will now be required)—and therein rests the principal merit of our present strategy.

IV.4   Eclipses of Oscillating Stars

The theory of the light changes of eclipsing binary systems, as outlined in the preceding chapter, has been based on a tacit assumption that (within the scheme of approximation adhered to so far) the surfaces of their components possess equilibrium forms appropriate for the prevalent field of force. The principal feature of such a model is the fact that, under such circumstances, any light changes exhibited by them—between minima as well as within eclipses—should be *symmetrical* with respect to the conjunctions. Asymmetry arising from orbital eccentricity (if any) is well understood, and generally very small (cf., Sec. IV-4 of Kopal, 1979); and even smaller are similar effects arising from possible tidal lag (Kopal, 1978; pp.124–126). And yet—in spite of these facts—the observed light changes of only too many eclipsing variables exhibit asymmetries which cannot have anything to do with the above-mentioned causes. Moreover, these are encountered in the sky too often to be attributed to some rare or unusual cause; and (as often as not) the sense of the asymmetry is found to fluctuate with the time—the well-known eclipsing system of U Cephei represents perhaps the best-known example of the case in point.

In recent years, such asymmetries have frequently been attributed to the occurrence of "starspots", distributed non-uniformly over the surfaces of rotating components, postulated (heuristically) to account for the observed facts. Such a hypothesis is, of course, neither new nor original (for some comments on its history cf., e.g., Kopal, 1982d); and its fuller discussion has been left wholly outside the scope of the present monograph, because of the fact that it represents a mere *"ad hoc"* hypothesis, incapable (so far) of any independent verification. Besides, it has already been amply proved by our astronomical ancestors (Bruns, 1882; Russell, 1906) that observations of the light changes arising allegedly from this cause cannot provide any *unique* information about the size or distribution of hypothetical spots over stellar surfaces; and, in consequence, any elaboration of such a hypothesis is, therefore, incapable of providing any positive results.

On the other hand, cumulative efforts to understand the physical causes of stellar variability have led, since the 1920's, to a realization that the principal cause of physical variability of the stars above the Main Sequence is *oscillations* of effective stellar photospheres, caused by their instability at certain stages of their evolution.

As is well known today, so long as the stars (including the components of binary systems—be these close or wide) derive their energy from a transmutation of hydrogen into helium, the very high thermal stability of such a process is sufficient to cause them to shine with constant light, and change their external characteristics only on the "nuclear" time-scale (too slow to disclose any such changes within centuries or millenia); the minor periodic variability of the Main-Sequence stars of the type of β CMa or β Cep being due to special causes.

However, for most stars which have evolved away from the Main Sequence, and whose evolution unrolls alternatively on the fast nuclear (helium) and dynamical

(Kelvin) time-scales, this will, in general, no longer be the case. In particular, certain domains are known to exist in the HR-diagram above the Main Sequence ("instability strips") where any star entering them in the course of its evolution is bound to exhibit physical variability of a particular type (cepheids, long-period variables, etc.); and the question arises as to the relevance of such phenomena also in double-star astronomy. Indeed, no reason is known which would restrict such a behaviour to single stars only; the latter are certainly not immune to it! In binary stars the components are, to be sure, gravitationally coupled; and this coupling may affect also their internal structure. In the deep interiors of the stars—where their nuclear energy is almost exclusively produced—the effects of gravitational coupling on internal structure are likely to be negligible; and any fluctuations which may occur there should be entirely unrelated with the orbital period of the system. The photospheric "gating effects", responsible for the variability of the cepheids (see, e.g., Zhevakin, 1953, 1954) may be affected by distortion to a much greater extent; and this should be especially true of "contact" components of the semi-detached binary systems. In such systems, a synchronization (or near-synchronization) of physical variability with orbital motion is more likely to occur; though a more specific investigation of this problem is still conspicuous by its absence. But whatever may turn out to be the case, a new class of observable phenomena may arise in this way which invites attention; and to these we wish to draw the reader's attention in the present section.

If the variable (oscillating?) component of an eclipsing pair happens to be the more luminous of the two—i.e., the "primary"—its variability may remain discernible throughout the entire orbital cycle—between minima as well as within eclipses; the eclipsing systems AB Cas or Y Cam (exhibiting also variability of δ Sct-type) may be regarded as typical examples of such a situation (others being AI Hya, EN Lac and no doubt others). On the other hand, the oscillating component may happen to be a secondary which is too faint to influence the combined light of the system to a detectable extent.

However, even low luminosity cannot prevent a disclosure of the secondary (eclipsing) component's physical variability if the latter is accompanied with a change of size or shape of its occulting disc within eclipses. Such a situation may well be expected to occur among the typical semi-detached Algol systems, in which a Main-Sequence primary is attended by an evolved mate, occupying a position in the HR-diagram where a star should be expected to oscillate if it were single. Algol itself shows no photometric evidence of instability of either component; but recent observations of fluctuating asymmetry of the minima of U Cep by Olson (1976a,b) strongly suggests that the eclipsing disc of the secondary component may indeed behave in such a manner; and the same is true of U Sge (cf. Olson, 1982), whose light minima due to total eclipses indicate that the secondary's shadow may be changing in size and (or) shape in the course of the minima. If so, the theory outlined in this section should demonstrate the manner in which the variations in shape of the oscillating components of close binary systems can affect the light changes exhibited within eclipses. The inverse problem (namely how to deduce such phenomena from the observations) will, however, be postponed for discussion till the next chapter.

For the time being, consider an Algol-like close eclipsing system consisting of two components which may differ considerably in mass; and in which, as a result, the primary (brighter and more massive) star can retain a nearly spherical form, while its companion has already evolved away from the Main Sequence to become a subgiant. This is not the place in which to discuss why or how this can come about; suffice it to state that such systems are known to exist in large numbers; and that their secondary components occupy the loci in the HR-diagram where the stars should be expected to become physical variables. Should they indeed do so, then photometric observations of the eclipse phenomena exhibited by them can open a new way for studying several aspects of physical variables in a way which would be inaccessible to us if the respective star were single.

For the present, let us assume that the loss of light l in an eclipsing system within minima can be expressed as

$$1 - l = (\alpha + f_1 + f_2)L_1 , \tag{4.1}$$

where L_1 denotes the fractional luminosity of the star undergoing eclipse, and the symbols α, $f_{1,2}$ have been defined by Eqs. (1.10) of Chapter II, and (2.20) with (3.30) of Chapter IV. If the primary component were spherical, $f_1 = 0$ identically; and if it shines with constant light, $L_1 = $ constant.

Let, in what follows, the function f_2 as given by Eq. (2.20) be restricted to terms factored by the amplitude $w_2^{(2)} \equiv \Delta_2 (m_2/m_1)r_2^3$ of second tidal harmonic only (a generalization of the procedure to harmonics of higher orders would offer no problem). If, moreover, the fractional radius r_2 of the secondary component is not constant, but becomes a function of the time—oscillating periodically in the time t about its mean value $(r_2)_0$ as $\exp i\sigma_m(t - t_0)$ with the frequency σ_m (depending only on the structure of the star's deep interior)—rendering

$$r_2 = (r_2)_0 + \delta r_2 ; \tag{4.2}$$

and δl stands for the variations of light due to δr_2; a partial differentiation of Eq. (4.1) with respect to r_2 discloses that, to a linear approximation,

$$\delta l = L_1 \sum_{n=0}^{N} C^{(n)} \frac{\partial}{\partial r_2} \left\{ \alpha_n^0 + \right.$$

$$+ \left(\frac{r_2}{r_1} \right)^{n+2} \left[3l_2^2 I_{-1,n}^2 - I_{-1,n}^0 \right] w_2^{(2)} +$$

$$\left. + \cdots \right\} = \tag{4.3}$$

$$= \frac{L_1}{r_2} \sum_{n=0}^{N} C^{(n)} \left(\frac{r_2}{r_1} \right)^{n+2} \left\{ 2I_{-1,n}^0 + \right.$$

$$+ \left[n + 5 + r_2 \frac{\partial}{\partial r_2} \right] \left[3l_2^2 I_{-1,n}^2 - I_{-1,n}^0 \right] w_2^{(2)} + \cdots \right\} \delta r_2,$$

where the term factored by $n + 5$ arises from the differentiation of the ratio $(r_2/r_1)^{n+2}$ and from the presence of r_2^3 in $w_2^{(2)}$. Moreover, it should be stressed that the phase t_0 in δr_2 need not be correlated in any way with the moments of the conjunctions (i.e.,the times of the minima of light).

A differentiation of $I_{-1,n}^0$ with respect to r_2 offers no difficulty: Equation (2.42) of Chapter II discloses that

$$\frac{\partial I_{-1,n}^0}{\partial r_2} = \frac{1}{2} \frac{\partial}{\partial r_2} \left\{ r_2 \left(\frac{r_1}{r_2} \right)^{n+2} \frac{\partial \alpha_n^0}{\partial r_2} \right\} =$$

$$= \frac{1}{2} \left(\frac{r_1}{r_2} \right)^{n+2} \left\{ r_2 \frac{\partial^2 \alpha_n^0}{\partial r_2^2} - (n+1) \frac{\partial \alpha_n^0}{\partial r_2} \right\}, \tag{4.4}$$

which combined with (2.77) of the same chapter yields

$$2 \left(\frac{r_2}{r_1} \right)^{n+2} \frac{\partial I_{-1,n}^0}{\partial r_2} = \frac{n}{2} \left\{ \frac{\delta^2 - r_1^2 - r_2^2}{r_1^2} \frac{\partial \alpha_{n-2}^0}{\partial r_2} - \frac{\partial \alpha_n^0}{\partial r_2} \right\} \tag{4.5}$$

or

$$r_2 \frac{\partial I_{-1,n}^0}{\partial r_2} = \frac{n}{2} \left\{ \frac{\delta^2 - r_1^2 - r_2^2}{r_2^2} I_{-1,n-2}^0 - I_{-1,n}^0 \right\} \tag{4.6}$$

for $n > 0$; while if $n = 0$,

$$r_2 \frac{\partial I_{-1,0}^0}{\partial r_2} = \frac{\delta^2 - r_1^2}{2 \delta r_2 \sqrt{1 - \mu^2}}, \tag{4.7}$$

where μ continues to be given by Eq. (2.27) of Chapter II.

Turning next to the eclipse function $I_{-1,0}^2$, let us return to Equations (2.24) and (2.28) of Chapter II: replacing in the latter $\gamma - 2$ by γ and inserting its left-hand side in (2.25); by use of (2.24) the latter yields the recursion relation

$$(\gamma + 2) \delta I_{\beta,\gamma}^{m+2} = -(\beta + m + 2) r_2 I_{\beta,\gamma+2}^{m+1} +$$
$$+ (\gamma + 2) \delta I_{\beta,\gamma}^m +$$
$$+ m r_2 I_{\beta,\gamma+2}^{m-1}, \tag{4.8}$$

which for $m = 0$, $\beta = -1$ and $\gamma = n$ yields

$$I_{-1,n}^2 = I_{-1,n}^0 - I_{1,n}^0 = I_{-1,0}^0 - \frac{r_1}{2\delta} \left(\frac{r_1}{r_2} \right)^{n+2} \alpha_n^1 \tag{4.9}$$

by Eq. (2.23) of Chapter II.

Let us, moreover, differentiate this foregoing equation partially with respect to r_2. The derivation of $I_{-1,n}^0$ is, however, known to us already from the preceding equation (4.6) or (4.7); while that of α_n^1 obtains (by means of Eq.(2.40) of Chapter II for $m = 0$) as

$$r_1 \frac{\partial \alpha_n^1}{\partial r_2} = \delta \frac{\partial \alpha_n^0}{\partial r_2} + r_2 \frac{\partial \alpha_n^0}{\partial \delta} =$$

$$= 2 \left(\frac{r_2}{r_1} \right)^{n+2} \left\{ \frac{\delta}{r_2} I_{-1,n}^0 - I_{-1,n}^1 \right\} \quad (4.10)$$

by (2.42) and (2.43) of Chapter II, so that

$$r_2 \frac{\partial I_{-1,0}^2}{\partial r_2} = n \left\{ \frac{\delta^2 - r_1^2 - r_2^2}{r_2^2} \right\} I_{-1,n-2}^0 -$$

$$- \frac{n+2}{2} I_{-1,n}^0 - \frac{r_2}{\delta} I_{-1,n}^1 . \quad (4.11)$$

By this all terms on the right-hand side of Equation (4.3)—representing photometric effects of the oscillations of the eclipsing star—have been explicitly evaluated in terms of the eclipse functions $I_{-1,n}^m$ ($m = 0$ and 1): those varying as $I_{-1,n}^0$ radial oscillations of the radius of shadow cylinder; and those factored by l_2^2, oscillations of non-radial type.

So far we have been concerned, in this chapter, with photometric effects arising from the oscillations—radial or non-radial—of the eclipsing component of a close binary system; while the light of the star undergoing eclipse is affected only by the secondary's progressing shadow. If, however, the primary also becomes a physical variable, possible oscillations in the size and shape of its apparent disc will likewise affect the combined light of the system; and as long as light perturbations arising from this source are small enough for their squares and higher powers to be negligible, those arising from physical activity of either component will remain additive.

In order to investigate similar changes due to physical activity of the primary component which undergoes eclipse, all we have to do is to insert in (4.1) from (3.31) for

$$f_1^{(n)} = - \left\{ a_2^{(0)} P_2(l_0) - a_2^{(1)} P_2^1(l_0) + a_2^{(2)} P_2^2(l_0) \right\} w_1^{(2)}, \quad (4.12)$$

where the $a_2^{(j)}$'s (for $j = 0, 1, 2$) are given by Eqs. (3.32)–(3.34). Moreover, by a resort to Eqs. (2.41)–(2.43) of Chapter II, a differentiation of (4.12) with respect to r_1 discloses that, if

$$r_1 \equiv (r_1)_0 + \delta r_1 , \quad (4.13)$$

a change of light δl caused by an oscillation δr_1 in radius of the star undergoing eclipse should be given by

$$\delta l = L_1 \sum_{n=0}^{N} C^{(n)} \frac{\partial}{\partial r_1} \left\{ \alpha_n^0 - \right.$$

$$- [a_2^{(0)}P_2(l_0) - a_2^{(1)}P_2^1(l_0) + a_2^{(2)}P_2^2(l_0)]w_1^2 +$$
$$+ \cdots\} \, \delta r_1 = \tag{4.14}$$

$$= -\frac{L_1}{r_1} \sum_{n=0}^{N} C^{(n)} \left\{ 2\mathfrak{S}^0_{-1,n} + \right.$$

$$+ \left(3 + r_1 \frac{\partial}{\partial r_1}\right) \left[a_2^{(0)}P_2(l_0) - a_2^{(1)}P_2^1(l_0) + \right.$$

$$\left. + a_2^{(2)}P_2^2(l_0)\right] w_1^{(2)} + \cdots \right\} \delta r_1$$

where, by Eq. (2.41) of Chapter II,

$$r_1 \frac{\partial a_2^{(0)}}{\partial r_1} = -3(\beta_2 - n)\mathfrak{S}^0_{-1,n+2} + (\beta_2 - 2n)\mathfrak{S}^0_{-1,n} - n\,\mathfrak{S}^0_{-1,n-2}, \tag{4.15}$$

$$r_1 \frac{\partial a_2^{(1)}}{\partial r_1} = (\beta_2 - n)r_1 \frac{\partial \alpha_n^1}{\partial r_1} \tag{4.16}$$

and

$$r_1 \frac{\partial a_2^{(2)}}{\partial r_1} = \frac{r_1}{2} \frac{\partial}{\partial r_1} \left\{ \frac{\beta_2 - 3h - 5}{h+1} \left[\frac{a_1}{\delta}\alpha_{n+2}^1 - \mathfrak{S}^0_{-1,n+2}\right] + \right.$$

$$\left. + \left[\frac{a_1}{\delta}\alpha_n^1 - \mathfrak{S}^0_{-1,n}\right] \right\}. \tag{4.17}$$

In order to perform the differentiation on the right-hand sides of the preceding two equations, let us return once more to Chapter II and combine its Equations (2.23) and (2.43) to obtain

$$\alpha_n^1 = -\frac{r_1}{n+2} \frac{\partial \alpha_{n+2}^0}{\partial \delta}, \tag{4.18}$$

which on differentiation with respect to r_1 yields

$$\frac{\partial \alpha_n^1}{\partial r_1} = -\frac{1}{n+2} \left\{ \frac{\partial \alpha_{n+2}^0}{\partial \delta} + r_1 \frac{\partial^2 \alpha_{n+2}^0}{\partial r_1 \partial \delta}; \right\} =$$

$$= -\frac{1}{n+2} \frac{\partial}{\partial \delta} \left\{ \alpha_{n+2}^0 - (n+2)\alpha_n^0 + n\alpha_{n-2}^0 \right\} \tag{4.19}$$

by (2.17); so that

$$r_1 \frac{\partial a_2^{(1)}}{\partial r_1} = 2\frac{\beta_2 - n}{n+2} \left(\frac{r_2}{r_1}\right)^{n-1} \left\{ \left(\frac{r_2}{r_1}\right)^4 I_{-1,n+2}^1 - \right.$$

$$\left. - (n+2)\left(\frac{r_2}{r_1}\right)^2 I_{-1,n}^1 + nI_{-1,n-2}^1 \right\} \tag{4.20}$$

while, by (2.17),

$$r_1 \frac{\partial \mathfrak{F}^0_{-1,n}}{\partial r_1} = -\frac{r_1}{2} \left\{ (n+2) \frac{\partial \alpha^0_n}{\partial r_1} - n \frac{\partial \alpha^0_{n-2}}{\partial r_1} \right\} =$$

$$= (n+2)\mathfrak{F}^0_{-1,n} - n\,\mathfrak{F}^0_{-1,n-2} \, . \tag{4.21}$$

On insertion from (2.41) of Chapter II in Eq. (3.34) of the present chapter, we eventually find that

$$r_1 \frac{\partial a_2^{(2)}}{\partial r_1} = \frac{\beta_2 - 1 - 6(\nu - 1)}{\nu} \left\{ \left(\frac{r_2}{r_1}\right)^4 I^0_{1,n+2} - \right.$$

$$- \nu \left(\frac{r_2}{r_1}\right)^2 I^0_{1,n} - \frac{\nu(\nu-1)}{\nu+1} I^0_{1,n-2} -$$

$$\left. -(\nu+1)\mathfrak{F}^0_{-1,n+2} + \nu\,\mathfrak{F}^0_{-1,n} \right\} + \tag{4.22}$$

$$+ \left\{ \left(\frac{r_2}{r_1}\right)^{n-2} \left[\left(\frac{r_2}{r_1}\right)^4 I^0_{1,n} - \left(\frac{r_2}{r_1}\right)^2 (\nu-1)I^0_{1,n-2} + \right.\right.$$

$$\left.\left. + \frac{(\nu-1)(\nu-2)}{\nu} I^0_{1,n-4} \right] - \nu \mathfrak{F}^0_{-1,n} + (\nu-1)\mathfrak{F}^0_{-1,n-2} \right\},$$

where $\nu \equiv (n+2)/2$.

The foregoing equations permit us to formulate explicitly the photometric effects of oscillations about the figures associated with second-harmonic distortion specified by the amplitudes $w_1^{(2)}$; while those associated with $w_1^{(3)}$, $w_1^{(4)}$ and $v_1^{(2)}$ can be evaluated in a similar manner only at the expense of an increasing amount of algebraic work.

IV.5   Bibliographical Notes

The subject matter of Section III.2, concerned with the photometric effects of distortion of the shadow cylinder cast by the eclipsing component along the line of sight owes its origin to Kopal (1941), and was developed further in subsequent literature quoted in the text. For a treatment of similar effects caused by distortion of the stars undergoing eclipses see Kopal (1942a,b). The presentation of the subject as given in terms of Hankel transforms, in Secs. III.2 and 3, is, however, new and has not been previously published.

The same is, moreover, true of the photometric effects due to free oscillations of the components of eclipsing binary systems. For a good account of physical theories of radial oscillations of the stars cf., e.g., Cox (1980); for non-radial oscillations cf. Unno et al (1979). All this work (without exception) was, however, concerned with free oscillations about *spherical* forms of equilibrium, appropriate for single stars in space (for oscillations of rotating stars cf. Tassoul, 1978). However, free oscillations about *distorted* forms of equilibrium in close binary systems represent virgin ground in this part of our vineyard.

It is true that mathematical problems arising in this connection have recently been formulated by the present writer in terms of Clairaut coordinates (in which the gravitational potential plays the role of the radial coordinate) (cf. Kopal, 1989; sec. V.3); but their solution still largely remains a task for the future.

The urgency of this task is underlined by the fact that the occurrence of free oscillations of the components in close eclipsing systems such as U Cep or U Sge seem indicated by the observations (cf. Olson, 1976a,b for U Cep; 1982 for U Sge). The photometric consequences of such motions were discussed so far only in Kopal (1982a,b); but much work remains yet to be done before their interpretation can be placed on a more solid footing.

Chapter V

INVERSE PROBLEM: SOLUTION FOR ELEMENTS OF THE ECLIPSES

In the preceding three chapters of this book an outline has been given of a mathematical theory of the light changes of eclipsing binary systems, which should permit us to evaluate the instantaneous light of the system at any moment of its orbital cycle—between minima as well as within eclipses—in terms of the geometrical as well as physical elements of the system. These elements are, however, not known to us *a priori*; in point of fact, it is the aim of the investigator to deduce them from the observed light changes. This constitutes, in fact, a problem *inverse* to that treated in the preceding two chapters, and one to which we wish to address ourselves in this chapter.

Such a task can be aptly compared with the *decoding* of the messages which reach us in the form of regular variations of their light; and theoretical light curves discussed in Chapters II–IV furnish a physically acceptable *code* for such an undertaking. Photometric observations of such systems performed by different methods are, however, available to us (so far) only in discrete form; for a sequential registration of periodically-varying light signals has been forced upon us by the (relatively) very low frequency at which such signals are emitted. Since the orbital periods (corresponding to the "zero harmonic") of close binary systems are, on the average, of the order of 10^5 seconds, periodic information reaching us from them on the waves of light are reaching us at frequencies of the order of 10^{-5} Hertz—i.e., some 23–24 octaves below that of audible sound!

A solution for the geometrical elements r_1, r_2, j of such systems from discrete date obtained by different methods are possible in principle; but in practice they require some care. It is true that—like in several other branches of dynamical astronomy (in celestial mechanics, for instance)—three observations are theoretically sufficient for a specification of the three above-mentioned geometrical elements. However, the individual photometric measurement l as a function of the time are so inferior in accuracy to those of the positions of the celestial bodies in the sky that it is well-nigh impossible to do so in practice.

In more specific terms, the photometrically-measured values of the brightness $l(\psi)$ of the system at a phase angle ψ defined by Eqs. (1.26) of the preceding Chapter IV are subject to observational errors of seldom less than 1% of the measured quantity, while the positional measurements by current astrometric methods can easily attain a precision of one part in several millions of the quadrant. In consequence, the difficulties besetting an analysis of the eclipse

phenomena, based on even the best of the observations that can be made so
far, should be comparable with those confronting the computer or asteroidal or
cometary orbits if the positional measurements at his disposal were inaccurate
within half a degree! Even Kidinnu and his Chaldean contemporaries in the 5th
century B.C. using the naked eye could have done much better! In such a situation, the relatively low accuracy of the underlying data must be compensated
by a large number of observations; and these require statistical, rather than individual, treatment: in particular, the weight of the solution must be based on a
wide stretch of the data rather than a limited number of discrete points.

     And this is what we must do in an analysis of the light changes of eclipsing
binary systems. A statistical treatment of the data (based on least-squares principles) necessary to this end can, to be sure, be left out of our present discussion;
for this has already been done elsewhere (for the latest treatment of this subject
cf. Kopal, 1986, which need not be repeated in this place). In what follows we
shall, however, focus our attention on the analytical aspects of such a procedure;
and this we cannot do better than by introducing to the reader, in the next section,the concept of the "moments of the light curves" which, in what follows, are
going to constitute the cornerstone of our analysis.

V.1   Moments of the Light Curves: Spherical Case

In order to embark on this task, consider first an eclipsing system which consists of
two spherical stars, revolving around the common centre of gravity, and appearing
in projection on the celestial sphere as uniformly bright (or arbitrarily limb-
darkened) discs; and consider the *area* subtended by the corresponding light
curve $l(\psi)$ in the $l - \sin^{2m} \psi$ coordinates, as shown on the accompanying Figure
V.1. The areas between the lines $l = 1$, $\sin^{2m} \psi = 0$ and $l(\psi)$ can evidently be
expressed in the form of the integral

$$A_{2m} = \int_0^{\psi_1} (1 - l) d(\sin^{2m} \psi), \qquad (1.1)$$

where ψ_1 denotes the phase angle of the first contact of the eclipse. The quantities
A_{2m}—hereafter referred to as the *moments of the eclipses*—depend only on data
supplied directly by the observations; and their empirical values can be deduced
from them by methods to be described later in Section II of this chapter. For
the present let us merely stress that such moments incorporate *all* information
supplied by the observations; and should, therefore, be capable of furnishing by
inversion the elements of the eclipse giving rise to the observed light changes $l(\psi)$.

     In order to accomplish this, let us set out to establish first the theoretical
values of the moments A_{2m} corresponding to the light changes

$$l = (1 - \alpha)L_1 + L_2 , \qquad (1.2)$$

normalized so that the sum of the luminosities $L_{1,2}$ of the two components has

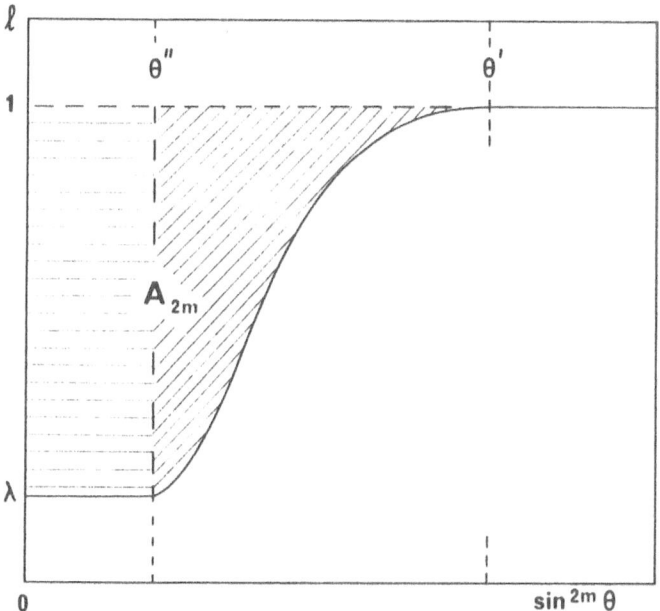

Figure V.1: Moments A_{2m} of the light curves.

been normalized to

$$L_1 + L_2 \; = \; 1 \; ; \tag{1.3}$$

and, in accordance with Eq. (1.10) of Chapter II,

$$\alpha \; = \; \sum_{n=0}^{N} C^{(n)} \, \alpha_n^0 \; . \tag{1.4}$$

Therefore, during eclipses

$$l \; = \; 1 = \alpha(\delta, k) L_1 \tag{1.5}$$

where, as before, $k \equiv r_1/r_2$; and—in accordance with Eq. (1.5) of Chapter II—the apparent separation δ of the projected centres of the two stars is given by the equation

$$\delta^2 \equiv \sin^2 \psi \, \sin^2 j + \cos^2 j \; = \; 1 - l_0^2 \, , \tag{1.6}$$

in which the angle j stands, as before, for the inclination of the orbital plane of our binary to the celestial sphere.

A differentiation of Equation (1.6) with respect to $\sin^2 \psi$ reveals that the element of integration on the right-hand side of Equation (1.1) can be expressed as

$$d(\sin^{2m} \psi) \; = \; m(\delta^2 - \delta_0^2)^{m-1}(1 - \delta_0^2)^{-m} d\delta^2, \tag{1.7}$$

where

$$\delta_0 \equiv \cos j \equiv n_0 ; \tag{1.8}$$

and, therefore,

$$A_{2m} = mL_1 \csc^{2m} j \int_{\delta_0^2}^{\delta_1^2} (\delta^2 - \delta_0^2)^{m-1} \alpha \, d\delta^2 =$$

$$= mL_1 \csc^{2m} j \sum_{n=0}^{N} C^{(n)} \int_{\delta_0^2}^{\delta_1^2} (\delta^2 - \delta_0^2)^{m-1} \alpha_n^0 \, d\delta^2 , \tag{1.9}$$

in which

$$\delta_{1,2} = |\, r_1 \pm r_2 \,| \tag{1.10}$$

signifies the first and second contact of the eclipse; such that if $\delta_0 > |\, r_1 - r_2 \,|$ the eclipse is partial; while in the converse case it is total for $\delta_0 < r_2 - r_1$ or annular if $\delta_0 < r_1 - r_2$.

Moreover, in accordance with Eq. (1.32) of Chapter III,

$$\alpha_n^0 = 2^{\nu+1} \Gamma(\nu) \int_0^{\infty} (2\pi q r_1)^{-\nu} J_\nu(2\pi q r_1) \times$$

$$\times J_1(2\pi q r_2) J_0(2\pi q \delta) d(2\pi q r_2); \tag{1.11}$$

so that

$$A_{2m} = mL_1 \csc^{2m} j \sum_{n=0}^{N} C^{(n)} 2^{\nu+1} \Gamma(\nu) \times$$

$$\times \int_0^{\infty} (2\pi r_1)^{-\nu} J_\nu(2\pi q r_1) J_1(2\pi q r_2) \left\{ \int_{\delta_0}^{\delta_1} (\delta^2 - \delta_0^2)^{m-1} \times \right.$$

$$\left. \times J_0(2\pi q \delta) \delta \, d\delta \right\} d(2\pi q r_2). \tag{1.12}$$

The reader should note that in none of the foregoing expressions (1.1)–(1.12) the finiteness of the moments A_{2m} requires that the constants $m \geq 0$ and $n \geq -1$; but neither m nor $n \equiv 2(\nu - 1)$ need be an integer; though in what follows we shall restrict our attention primarily to integral values of m.

By successive application of the differential recursion formula

$$(2\pi q)\delta^j J_{j-1}(2\pi q \delta) = \frac{d}{d\delta} \{\delta^j J_j(2\pi q \delta)\}, \tag{1.13}$$

valid for Bessel functions of arbitrary order, we readily establish that, for $m = 1$,

$$\int_{\delta_0}^{\delta_1} J_0(2\pi q \delta)\delta \, d\delta = \{h_1 x^{-1} J_1(h_1 x) - h_0 x^{-1} J_1(h_0 x)\} r_2^2; \tag{1.14}$$

while for $m = 2$,

$$\int_{\delta_0}^{\delta_1} (\delta^2 - \delta_0^2) J_0(2\pi q \delta)\delta \, d\delta = \{h_1^2 - h_0^2)(h_1/x) J_1(h_1 x) - 2(h_1/x)^2 J_2(h_1 x) +$$

$$+ 2(h_0/x)^2 J_2(h_0 x)\} r_2^4, \tag{1.15}$$

for $m = 3$,

$$\int_{\delta_0}^{\delta_1} (\delta^2 - \delta_0^2)^2 J_0(2\pi q \delta) \delta \, d\delta = \{ (h_1^2 - h_0^2)^2 (h_1/x) J_1(h_1 x) - -$$
$$- 4(h_1^2 - h_0^2)(j_1/x)^2 J_2(h_1 x) +$$
$$+ 8(h_1/x)^3 J_3(h_1 x) -$$
$$- 8(h_0/x)^3 J_3(h_0 x) \} \, r_2^6 \qquad (1.16)$$

where (cf. Eq. 1.33 of Chapter III) $x \equiv 2\pi q r_2$ and

$$h_1 = \frac{\delta_1}{r_2} = \frac{r_1 + r_2}{r_2} = 1 + k, \qquad (1.17)$$

while

$$h_0 = \frac{\delta_0}{r_2} = \frac{\cos j}{r_2}. \qquad (1.18)$$

in accordance with Eqs. (1.35) of Chapter III.

If, furthermore, we abbreviate

$$2^{\nu+1} \Gamma(\nu) h_j^m \int_0^\infty \frac{J_\nu(kx)}{(kx)^\nu} J_1(x) \frac{J_m(h_j x)}{x^m} dx \equiv \Pi_m^{(\nu)}(h_j), \qquad (1.19)$$

where $j = 0$ or 1, it follows from Equation (1.12) that

$$A_2 = L_1 r_2^2 \csc^2 j \sum_{l=0}^{\infty} C^{(l)} \{ \Pi_1^{(\nu)}(h_1) - \Pi_1^{(\nu)}(h_0) \}, \qquad (1.20)$$

$$A_4 = 2 L_1 r_2^4 \csc^4 j \sum_{l=0}^{\infty} C^{(l)} \left\{ (h_1^2 - h_0^2) \Pi_1^{(\nu)}(h_1) - \right.$$
$$\left. - 2 \Pi_2^{(\nu)}(h_1) + 2 \Pi_2^{(\nu)}(h_0) \right\}, \qquad (1.21)$$

$$A_6 = 3 L_1 r_2^6 \csc^6 j \sum_{l=0}^{\infty} C^{(l)} \left\{ (h_1^2 - h_0^2)^2 \Pi_1^{(\nu)}(h_1) - \right.$$
$$- 4(h_1^2 - h_0^2) \Pi_2^{(\nu)}(h_1) + 8 \Pi_3^{(\nu)}(h_1) -$$
$$\left. - 8 \Pi_3^{(\nu)}(h_0) \right\}, \qquad (1.22)$$

etc.

In order to evaluate the Π-integrals on the right-hand sides of the preceding equations, consider first those with the argument h_1, referring to the first contact of the eclipse. Since, at that moment, $h_1 = 1 + k$, the use of Bailey's equation (2.6) of Chapter III discloses that

$$\Pi_m^{(\nu)}(h_1) = \frac{1}{\nu \Gamma(m)} \left(\frac{h_1^2}{2} \right)^{m-1} F^{(4)}(1 - m, 1; 2, \nu + 1; h_1^{-2}, k^2/h_1^2) \qquad (1.23)$$

for *any* type of eclipse, where $m = 1, 2, 3, \ldots$ and, therefore, Appell's generalized hypergeometric series $F^{(4)}$ on the right-hand side of (1.23), as given by Eq. (2.8) of Chapter III, reduces to a polynomial—regardless of the degree of the law (1.2) of limb-darkening.

It is only the Π's with the argument h_0 (corresponding to the moment of maximum eclipse) which will be different for different types of eclipse; and the following three cases must be distinguished. If the eclipse is *total*, $\delta < r_2 - r_1$ and, therefore, $1 > h_0 + k$. In such a case, Bailey's formula (2.6) continues to apply (the series $F^{(4)}$ reducing to its first term) and yields

$$\Pi_m^{(\nu)}(h_0) = \frac{2}{m!\nu}\left(\frac{h_0^2}{2}\right)^m ; \tag{1.24}$$

which on insertion to (1.20)–(1.22) together with (1.23) verifies the expressions

$$A_2 = L_1 \overline{C}_3 , \tag{1.25}$$

$$A_4 = L_1(\overline{C}_3^2 + \overline{C}_2^2) , \tag{1.26}$$

$$A_6 = L_1(\overline{C}_3^3 + 3\overline{C}_2^2\overline{C}_3 + \overline{C}_1\overline{C}_2^2) , \tag{1.27}$$

where

$$\overline{C}_3 = (r_2^2 \csc^2 j - \cot^2 j)\sum_{n=0}^{N}\frac{1!C^{(n)}}{\nu} , \tag{1.28}$$

$$\overline{C}_2^2 = r_1^2 r_2^2 \csc^4 j \sum_{n=0}^{N}\frac{2!C^{(n)}}{\nu(\nu+1)} , \tag{1.29}$$

$$\overline{C}_1\overline{C}_2^2 = r_1^4 r_2^2 \csc^6 j \sum_{n=0}^{N}\frac{3!C^{(n)}}{\nu(\nu+1)(\nu+2)} . \tag{1.30}$$

The reader may note that, of the constants $\overline{C}_{1,2,3}$ introduced in t he preceding equations (1.28)–(1.30), the first two ($\overline{C}_{1,2}$) are bound to be positive regardless of the type of eclipse; but $\overline{C}_3$ may be positive or negative depending on whether $r_2 \gtrless \cos i$. Moreover, since (consistent with Eqs. (1.11)–(1.12) of Chapter II),

$$\sum_{n=0}^{N}\frac{1!C^{(n)}}{\nu} = \sum_{n=0}^{N}\frac{2C^{(n)}}{n+2} = 1 , \tag{1.31}$$

it follows that

$$\overline{C} \equiv C_3 \tag{1.32}$$

i.e., that *the quantity C_3 is invariant with respect to the limb-darkening*—a fact not true of $C_{1,2}$.

If, next, the eclipse happens to be *annular*—i.e., if $r_1 > \delta_0 + r_2$ and, therefore, $k > h_0 + 1$, Bailey's equation (2.6) of Chapter III continues to hold good; and its application to (1.23) yields

$$\Pi_m^{(\nu)}(h_0) = \frac{2}{m! k^2} \left(\frac{h_0^2}{2}\right)^m F^{(4)}(1 - \nu, 1; 2, m + 1; k^2, h_0^2/k^2), \qquad (1.33)$$

as a result of which

$$A_2 = L_1 \overline{C}_3 + L_1 \cot^2 j \left\{1 - (r_2/r_1)^2 \sum_{n=0}^{\infty} C^{(n)} \times \right.$$
$$\left. \times F^{(4)}(1 - \nu, 1; 2, 2; r_2^2/r_1^2, \delta_0^2/r_1^2)\right\}, \qquad (1.34)$$

$$A_4 = L_1 (\overline{C}_3^2 + \overline{C}_2^2) - L_1 \cot^4 j \left\{1 - (r_2/r_1)^2 \sum_{n=0}^{\infty} C^{(n)} \times \right.$$
$$\left. \times F^{(4)}(1 - \nu, 1; 2, 3; r_2^2/r_1^2, \delta_0^2/r_1^2)\right\}, \qquad (1.35)$$

$$A_6 = L_1 (\overline{C}_3^3 + 3\overline{C}_2^2 \overline{C}_3 + \overline{C}_1 \overline{C}_2^2) +$$
$$+ L_1 \cot^6 j \left\{1 - (r_2/r_1)^2 \sum_{n=0}^{\infty} C^{(n)} \times \right.$$
$$\left. \times F^{(4)}(1 - \nu, 1; 2, 4; r_2^2/r_1^2, \delta_0^2/r_1^2)\right\}, \qquad (1.36)$$

can again be expressed in terms of Appell's $F^{(4)}$-series, which reduce to polynomials provided that $1 - \nu$ happens to be a negative integer (which it will be for even values of n).

V.2 Moments of the Light Curves: General Properties

Should, however, $r_1 + r_2 > \delta_0 > |r_1 - r_2|$, the eclipse becomes *partial*; but if so, none of the three parameters h_0, k and 1 becomes larger than the sum of the two others. As a result, the condition (2.9) of the validity of Bailey's theorem (2.6) in Chapter III cannot be met for that theorem to help us evaluate the $\Pi_m^{(\nu)}(h_0)$'s as defined by Eq. (1.19); and, as a result, the moments A_{2m} as given by (1.9) cannot be determined by means of (1.12) in a finite number of terms.

However, we may recall from Section III.2 that, by a resort to Bateman's expansion (2.28) we succeeded in constructing for the function α_n^0 an infinite expansion of the form (2.38), which can be inserted in the integrand on the r.h.s. of Eq. (1.9) of the present chapter to evaluate a corresponding expansion for A_{2m}: in doing so we find that

$$A_{2m} = \frac{m L_1}{(1 - n_0^2)^m} \sum_{n=0}^{N} \frac{C^{(n)}}{(\nu)_2} \sum_{j=0}^{\infty} (j + 1)(\nu + j + 1) \times$$

$$\times (\nu + 2j + 2)\{bG_j(\nu + 2, 2; b)\}^2 \int_{\delta_0^2}^{\delta_1^2} (\delta^2 - \delta_0^2)^{m-1} \times$$

$$\times (1 - c^2)^{\nu+1} G_j(\nu + 2, \nu + 2; 1 - c^2) d\delta^2. \tag{2.1}$$

In order to normalize the limits of integration on the right-hand side of the preceding equation from (δ_0^2, δ_1^2) to $(0, 1)$, we introduce a new variable u defined by Equation (4.16) of Chapter III as

$$u = \frac{\delta^2 - \delta_0^2}{\delta_1^2 - \delta_0^2} = \frac{c^2 - c_0^2}{1 - c_0^2} \tag{2.2}$$

and integrate the right-hand side of (2.1) term-by-term: in doing so we find that

$$A_{2m} = L_1(b \sin^m \psi_1)^2 \sum_{n=0}^{N} \frac{C^{(n)}}{\nu} B(\nu + 1, m + 1) \times$$

$$\times (1 - c_0^2)^{\nu+1} \sum_{j=0}^{\infty} (j + 1)(\nu + j + 1)(\nu + 2j + 2) \times$$

$$\times \{G_j(\nu + 2, 2; b)\}^2 G_j(\nu + 2, \nu + m + 2; 1 - c_0^2), \tag{2.3}$$

where ψ_1 stands (as before) for the phase angle of first contact of the eclipse; B, for the (complete) beta-function of its arguments (i.e., a numerical factor); and the parameters b and c_0 continue to be defined by Eqs. (2.4)–(2.5) of Chapter III as

$$b = \frac{r_2}{r_1 + r_2} \quad \text{and} \quad c_0 = \frac{n_0}{r_1 + r_2}. \tag{2.4}$$

The principal virtue of the above equation (2.3) is again its extreme generality: for it holds good equally for *any* type of eclipse—be these total, annular or partial (occultation or transits); and for any (non-negative) values of m—be it integral, fractional or irrational. What happens when $m = 0$? Since

$$\lim_{m \to 0} d(\sin^{2m} \psi) = 0 \text{ for } \psi > 0,$$
$$= 1 \text{ for } \psi = 0, \tag{2.5}$$

Equation (1.1) discloses that A_0 no longer represents an area subtended by the light curve, but its ordinate at $\psi = 0$. In other words, the quantity A_0 is identical with the maximum fractional loss of light $1 - \lambda$ suffered by the system at zero phase. For total eclipses, therefore, we evidently have

$$A_0 = L_1 ; \tag{2.6}$$

while for partial (or annular) eclipses,

$$A_0 = L_1 \alpha_0 , \tag{2.7}$$

where

$$\alpha_0 \equiv \sum_{n=0}^{N} \alpha_n^0(h_0, k) \tag{2.8}$$

stands for the *maximum obscuration* of the eclipsed star.

The moments A_{2m} $(m \geq 0)$ of the light curves possess also other general properties of considerable interest which should be pointed out in view of their practical applications. As is well known, the functions α_n^0 (and, therefore, any of their linear combinations α as defined by Eq. 1.4) satisfy Euler's equation

$$r_1^2 \frac{\partial \alpha}{\partial r_1^2} + r_2^2 \frac{\partial \alpha}{\partial r_2^2} + \delta^2 \frac{\partial \alpha}{\partial \delta^2} = 0 \qquad (2.9)$$

for homogeneous functions of zero degree; and, therefore,

$$r_1^2 \frac{\partial A_0}{\partial r_1^2} + r_2^2 \frac{\partial A_0}{\partial r_2^2} + \delta_0^2 \frac{\partial A_0}{\partial \delta_0^2} = 0 \qquad (2.10)$$

by (2.7).

The trio $r_{1,2}$ and $\delta_0 \equiv \cos j$ do not, however, represent any unique choice of such parameters. For consider an alternative set

$$C_1 = r_1^2 \csc^2 j , \qquad (2.11)$$

$$C_2 = r_1 r_2 \csc^2 j , \qquad (2.12)$$

$$C_3 = r_2^2 \csc^2 j - \cot^2 j, \qquad (2.13)$$

to which the $\overline{C}_j$'s as given by Eqs. (1.28)–(1.30) reduce for uniformly-bright discs. Consider now the differential operators rewritten in terms of these C_j's: namely,

$$C_1 \frac{\partial}{\partial C_1} \equiv (1 + r_2^2) r_1^2 \frac{\partial}{\partial r_1^2} - (1 - r_2^2) r_2^2 \frac{\partial}{\partial r_2^2} +$$
$$+ r_2^2 \sin^2 j \frac{\partial}{\partial \sin^2 j}, \qquad (2.14)$$

$$C_2 \frac{\partial}{\partial C_2} \equiv -2 r_1^2 r_2^2 \frac{\partial}{\partial r_1^2} + 2(1 - r_2^2) r_2^2 \frac{\partial}{\partial r_2^2} -$$
$$- 2 r_2^2 \sin^2 j \frac{\partial}{\partial \sin^2 j} \qquad (2.15)$$

and

$$\frac{\partial}{\partial C_3} \equiv \sin^2 j \left\{ r_1^2 \frac{\partial}{\partial r_1^2} + r_2^2 \frac{\partial}{\partial r_2^2} + \sin^2 j \frac{\partial}{\partial \sin^2 j} \right\}, \qquad (2.16)$$

which stands apart from the preceding two operators by virtue of the fact that it is symmetrical with respect to $r_{1,2}$ and is, therefore, valid for any type of eclipse (be these occultations or transits). A sum of Eqs. (2.14)–(2.16) reduces then to

$$C_1 \frac{\partial}{\partial C_1} + C_2 \frac{\partial}{\partial C_2} + C_3 \frac{\partial}{\partial C_3} =$$
$$= \sin^2 j \left\{ r_1^2 \frac{\partial}{\partial r_1^2} + r_2^2 \frac{\partial}{\partial r_2^2} + \delta_0^2 \frac{\partial}{\partial \delta_0^2} \right\}, \qquad (2.17)$$

which combined with (2.10) discloses that

$$C_1 \frac{\partial A_0}{\partial C_1} + C_2 \frac{\partial A_0}{\partial C_2} + C_3 \frac{\partial A_0}{\partial C_3} = 0. \tag{2.18}$$

Let us, however, return to Equation (2.16) above, and differentiate (1.1) with respect to C_3 behind the integral sign. Since the loss of light $1 - l$ commences at the moment of first contact, $f(\psi_1) = 0$ and, consequently,

$$\frac{\partial A_{2m}}{\partial C_3} = L_1 \int_0^{\psi_1} \frac{\partial \alpha}{\partial C_3} d(\sin^{2m} \psi); \tag{2.19}$$

or, changing over from ψ to δ as the variable of integration we can rewrite (2.19) as

$$\frac{\partial A_{2m}}{\partial C_3} = -mL_1 \csc^{2(m-1)} j \int_{\delta_0^2}^{\delta_1^2} \left\{ \delta^2 \frac{\partial \alpha}{\partial \delta^2} - \delta_0^2 \frac{\partial \alpha}{\partial \delta_0^2} + \frac{\partial \alpha}{\partial \delta_0^2} \right\} (\delta^2 - \delta_0^2)^{m-1} d\delta^2. \tag{2.20}$$

On the other hand, by lowering the index of A_{2m} from m to $m - 1$, it also follows from (1.1) that

$$A_{2(m-1)} = L_1 \int_0^{\psi_1} d(\sin^{2(m-1)} \psi) =$$

$$= (m - 1) L_1 \csc^{2(m-1)} j \int_{\delta_0^2}^{\delta_1^2} (\delta^2 - \delta_0^2)^{m-2} \alpha \, d\delta^2 =$$

$$= -L_1 \csc^{2(m-1)} j \int_{\delta_0^2}^{\delta_1^2} (\delta^2 - \delta_0^2)^{m-1} \frac{\partial \alpha}{\partial \delta^2} d\delta^2 \tag{2.21}$$

by partial integration; so that, by a subtraction of (2.20) and (2.21), we obtain

$$\frac{\partial A_{2m}}{\partial C_3} - mA_{2(m-1)} = mL_1 \csc^{2(m-1)} j \times$$

$$\times \int_{\delta_0^2}^{\delta_1^2} \left\{ (1 - \delta^2) \frac{\partial \alpha}{\partial \delta^2} - (1 - \delta_0^2) \frac{\partial \alpha}{\partial \delta_0^2} \right\} (\delta^2 - \delta_0^2)^{m-1} d\delta^2. \tag{2.22}$$

From Equation (1.5) it follows, however, that

$$(1 - \delta^2) \frac{\partial \alpha}{\partial \delta^2} = \frac{\partial \alpha}{\partial \delta^2} \sin^2 j \cos^2 \psi \equiv l_0 \frac{\partial \alpha}{\partial \delta^2}. \tag{2.23}$$

Therefore, the difference in curly brackets of Equation (2.22) will be zero in the whole range of $\delta_0 \le \delta \le \delta_1$—a fact which is sufficient to annihilate the entire right-hand side of this equation; and to leave us with an outcome disclosing that

$$\frac{\partial A_{2m}}{\partial C_3} = mA_{2(m-1)}, \tag{2.24}$$

which represents a fundamental differential recursion formula satisfied by our moments A_{2m} of the light curves. In the case of total eclipses, the existence of the foregoing relation is readily verified by a differentiation of Eqs. (1.25)–(1.27) giving the moments $A_{2,4,6}$; and, in the general case, by a differentiation of Eq. (2.3) when advantage is taken of the fact that

$$\frac{\partial}{\partial(1 - c_0^2)}\{(1 - c_0^2)^{m+\nu+1}G_j(\nu + 1, \nu + m + 2; 1 - c_0^2)\} =$$
$$= (m + \nu + 1)(1 - c_0^2)^{\nu+m}G_j(\nu + 2, \nu + m + 1; 1 - c_0^2). \qquad (2.25)$$

Before we proceed to discuss the significance of the remarkably simple relation (2.24), let us underline once more its generality. The only condition for its validity is, in fact, the requirement that the function $\alpha(r_1, r_2, \delta_0)$ be a homogeneous function of zero degree in its parameters! But this is bound to be true for any type of eclipse—occultation or transit; partial, total or annular—and any degree of limb-darkening. And whatever may be the case, the moments A_{2m} of the light for *any* (not necessarily integral!) value of $m \geq 0$ are found to belong to a class of functions satisfying the difference-differential equation (2.24), the general properties of which were studied first by Appell (1880) and, more recently, by Truesdell (1948).

Next, we should keep in mind that, for $m > 0$, *the moments A_{2m} are homogeneous functions of m-th degree*; and as such they satisfy Euler's differential equation

$$\left(C_1\frac{\partial}{\partial C_1} + C_2\frac{\partial}{\partial C_2} + C_3\frac{\partial}{\partial C_3}\right)^n A_{2m} = n!\binom{m}{n} A_{2m} \qquad (2.26)$$

for each integral value of $n \geq 0$; and for $n = 1$ it assumes the form

$$\left(C_1\frac{\partial}{\partial C_1} + C_2\frac{\partial}{\partial C_2} + C_3\frac{\partial}{\partial C_3}\right) A_{2m} = \qquad (2.27)$$
$$= \sin^2 j\left\{r_1^2\frac{\partial}{\partial r_1^2} + r_2^2\frac{\partial}{\partial r_2^2} + \delta_0^2\frac{\partial}{\partial \delta_0^2}\right\} A_{2m} = mA_{2m};$$

which combined with (2.24) discloses that

$$\left(r_1^2\frac{\partial}{\partial r_1^2} + r_2^2\frac{\partial}{\partial r_2^2}\right) A_{2m} = m(A_{2m} + A_{2(m-1)}) \cot^2 j. \qquad (2.28)$$

In conclusion, let us return to Eq. (2.24) and recall, from Appell's work, that if (in our present case)

$$A_{2m} = L_1 \sum_{j=0}^{m} \binom{m}{j} v_j C_3^{m-j} \qquad (2.29)$$

where the v_j's are polynomials in C_1 and C_2; and if

$$u(h) = \sum_{j=0}^{\infty} \frac{v_j}{j!} h^j , \qquad (2.30)$$

represent a function of a dummy variable in terms of the polynomials v_j, then

$$L_1 u(h) e^{hC_3} = \sum_{j=0}^{\infty} \frac{h^j}{j!} A_{2j},$$

(2.31)

which represents *generating function of the moments A_{2j} of the light curves* of even orders. Such functions can, in turn, be used to facilitate a solution for the elements of the eclipses in terms of the moments A_{2m} of their respective light curves; and to this task we shall turn our attention in section V-3; while the aim of the present section should be to demonstrate that the equations given so far to represent the moments A_{2m} as defined by (1.1) so far are neither necessary, nor unique.

In order to do so, let us recall that an expression for α_n^0 alternative to (2.38) of Chapter III can be obtained by a resort to a summation theorem for Bessel functions in the form

$$J_0(cy) = \left(\frac{2}{y}\right)^{1/2} \sum_{j=0}^{\infty} \left(2j + \frac{1}{2}\right) \frac{\Gamma(j + \frac{1}{2})}{\Gamma(j + 1)} P_{2j}(\sqrt{1 - c^2}) J_{2j+\frac{1}{2}}(y),$$

(2.32)

where the P_{2j}'s stand for the Legendre polynomials of even orders of the argument $\sqrt{1 - c^2}$.

On inserting this expansion on the r.h.s. of the preceding Equation (2.32) we find the latter to asume the form

$$\alpha_n^0 = 2^{\nu + \frac{1}{2}} \Gamma(\nu) a^{-\nu} b \sum_{j=0}^{\infty} (2j + \frac{1}{2}) \frac{\Gamma(j + \frac{1}{2})}{\Gamma(j + 1)} P_{2j}(\sqrt{1 - c^2}) \times$$

$$\times y^{-\nu - \frac{1}{2}} J_\nu(ay) J_1(by) J_{2j+\frac{1}{2}}(y) dy.$$

(2.33)

Since $a + b = 1$, Bailey's theorem (2.6) can help us once more to evaluate the integrals on the r.h.s. of the preceding equation; with the outcome disclosing that

$$\alpha_n^0 = \frac{b^2}{\nu} \sum_{j=0}^{\infty} (2j + \frac{1}{2}) F^{(4)}(-j + \frac{1}{2}, j + 1; \nu + 1,$$

$$2; a^2, b^2) P_{2j}(\sqrt{1 - c^2}),$$

(2.34)

where

$$P_{2j}(\sqrt{1 - c^2}) = \frac{(-1)^j (2j)!}{2^{2j} (j!)^2} \, {}_2F_1\left(-j, j + \frac{1}{2}; \frac{1}{2}; 1 - c^2\right) =$$

$$= P_{2j}(0) G_j\left(\frac{1}{2}, \frac{1}{2}; 1 - c^2\right).$$

(2.35)

If, furthermore, we employ the foregoing expansions (2.34)–(2.35) in place of (2.38), Chapter III, in the integrand of (1.1) we find that

$$A_{2m} = b^2 L_1 \sin^{2m} \psi_1 \sum_{n=0}^{N} \frac{C^{(n)}}{\nu} \sum_{j=0}^{\infty} (2j + \frac{1}{2}) P_{2j}(0) \times$$

$$\times F^{(4)}(-j + \frac{1}{2}, j + 1; \nu + 1, 2; a^2, b^2) \times$$

$$\times \, _3F_2(-j, j + \frac{1}{2}; 1; \frac{1}{2}, m + 1; 1 - c_0^2). \tag{2.36}$$

Or if again we resort to an expansion for α_n^0 of the form

$$\nu \alpha_n^0 = 2b^2 \sum_{j=0}^{\infty} j F^{(4)}(1 - j, 1 + j; 1 + \nu, 2; a^2, b^2) G_j(0, 1; c^2), \tag{2.37}$$

established by Demircan (1978), the corresponding expansion for the moments A_{2m} assume the form

$$A_{2m} = b^2 L_1 \sin^{2m} \psi_1 \sum_{n=0}^{N} \frac{C^{(n)}}{\nu} \sum_{j=0}^{\infty} (-1)^j (2j + 1) \times$$

$$\times F^{(4)}(-j, j + 1; \nu + 1, 2; a^2, b^2) G_j(1, m + 1; 1 - c_0^2). \tag{2.38}$$

Neither of the series on the right-hand sides of the foregoing equations (2.36) or (2.38) can be matched term-by-term with each other or with (2.31); but all converge to the same result at different rates in different domains of a and c_0 for which they are valid. Should, moreover, $a = 0$ (i.e., the occulting limb become a straight edge), when the fractional loss of light α_n^0 can be expressed in terms of the geometrical depth p of the eclipse by Equations (2.48) and (2.59) of Chapter III. Since, moreover, by the former

$$\sin^2 \psi = C_1 p^2 + C_2 p + C_3, \tag{2.39}$$

where the constants $C_{1,2,3}$ continue to be defined by Eqs. (2.11)–(2.13) earlier in this section, while the derivative of α_n^0 with respect to p is given by Eq. (2.58), Chapter III.

If so, a partial integration of Eq. (1.1) of this chapter yields

$$A_{2m} = L_1 \sum_{n=0}^{n} C^{(n)} \int_{1}^{0} \sin^{2m} \psi \, d\alpha_n^0 =$$

$$= L_1 \sum_{n=0}^{n} \frac{C^{(n)}}{\nu} \left\{ \frac{\Gamma(\nu + 1)}{\sqrt{\pi} \Gamma(\nu + \frac{1}{2})} \right\} \int_{-1}^{1} (1 - p^2)^{\nu - \frac{1}{2}} \times$$

$$\times (C_1 p^2 + C_2 p + C_3)^m \, dp, \tag{2.40}$$

which for $m = 1, 2, 3, \ldots$ yields

$$A_2 = C_2 \sum_{n=0}^{N} \frac{C^{(n)}}{(\nu)_2 B(\frac{1}{2}, \nu + \frac{3}{2})}, \tag{2.41}$$

$$A_4 = 2C_1 C_2 \sum_{n=0}^{N} \frac{C^{(n)}}{(\nu)_3 B(\frac{1}{2}, \nu + \frac{5}{2})} +$$

$$+ 2C_2 C_3 \sum_{n=0}^{N} \frac{C^{(n)}}{(\nu)_2 B(\frac{1}{2}, \nu + \frac{3}{2})}, \tag{2.42}$$

$$A_6 = 6C_1^2 C_2 \sum_{n=0}^{N} \frac{C^{(n)}}{(\nu)_4 B(\frac{1}{2}, \nu + \frac{7}{2})} +$$

$$+ C_2(C_2^2 + 6C_1 C_3 \sum_{n=0}^{N} \frac{C^{(n)}}{(\nu)_3 B(\frac{1}{2}, \nu + \frac{5}{2})} +$$

$$+ 3C_2 C_3^2 \sum_{n=0}^{N} \frac{C^{(n)}}{(\nu)_2 B(\frac{1}{2}, \nu + \frac{3}{2})}, \tag{2.43}$$

etc.

Our discussion of this topic should be concluded by an application of a similar technique to evaluate the moments A_{2m} with the aid of the Fourier expansion (4.25) of Chapter III. For it we integrate the right-hand side of Equation (1.1) of the present chapter by parts, the latter can be rewritten as[1]

$$A_{2\mu} \equiv L_1 \int_0^{\psi_1} \alpha \, d \sin^{2\mu} \psi = L_1 \int_0^{\alpha_0} \sin^{2\mu} \psi \, d\alpha \tag{2.44}$$

which by differentiating (4.25) of Chapter III we can rewrite as

$$A_{2\mu} = L_1 \sum_{n=0}^{n} C^{(n)} \sum_{m=1}^{\infty} m B_m^{(n)} \int_0^{\psi_1} \sin^{2\mu} \psi \sin m\psi \, d\psi \tag{2.45}$$

in which (since $\alpha = 0$ for $\frac{\pi}{2} \geq \psi \geq \psi_1$) the upper limit on the right-hand side for ψ can be replaced by $\psi_1 = 90°$; and the $B_m^{(n)}$'s continue to be given by Eqs. (4.26) or (4.27) of that chapter.

If so, however, by Eq. (4.29) of the same chapter

$$\int_0^{\pi/2} \sin^{2\mu} \psi \sin m\psi \, d\psi =$$

$$= m \int_0^{\pi/2} \sin^{2\mu+1} \psi \, {}_2F_1 \left(\frac{1+m}{2}, \frac{1-m}{2}; \frac{3}{2}; \sin^2 \psi \right) d\psi =$$

$$= \frac{m}{2} \int_0^1 x^\mu (1-x)^{-1/2} \, {}_2F_1 \left(\frac{1+m}{2}, \frac{1-m}{2}; \frac{3}{2}; x \right) dx \tag{2.46}$$

by $\sin^2 \psi \equiv x$; and can be evaluated in terms of a generalized hypergeometric series of unit radius (cf. Erdélyi *et al*, 1954; p.399(5)) to yield

$$\int_0^{\pi/2} \sin^{2\mu} \psi \sin m\psi \, d\psi = \frac{m}{2} B(m+1, \frac{1}{2}) \times$$

$$\times {}_3F_2 \left(\frac{1+m}{2}, \frac{1-m}{2}, \mu+1; \frac{3}{2}, \mu+\frac{3}{2}; 1 \right) \tag{2.47}$$

where m stands for zero or a positive integer; but μ may, but need not necessarily be so.

---

[1] In which the quantity m of (1.1) has been replaced by μ to underline the fact that the latter need not be an integer.

V.3 Inversion of the Moments: Spherical Stars

Should the eclipses of spherical stars giving rise to the observed light changes be *total*, the task of expressing the elements of such eclipses in terms of the moments of the respective eclipses becomes trivially simple: namely, if $\delta < r_2 - r_1$, $\alpha(\delta, k)$ $= 1$ in Eq. (1.5) for any value of k and, accordingly,

$$L_1 = 1 - \lambda, \tag{3.1}$$

where λ denotes the fractional luminosity of the system during totality (which can be ascertained directly from the observations). And once its numerical value has been established, that of $\overline{C}_3$ obtains readily from Eq. (1.25); $\overline{C}_2$ from Eq. (1.26) and $\overline{C}_1$ from (1.27) in that order.

Moreover, if the coefficients u_j of limb-darkening of the star undergoing eclikpse can be estimated from its spectral type (or otherwise), Eqs. (1.28)–(1.32) should permit us to convert the $\overline{C}_j$'s in the C_j's as defined by Eqs. (2.11)–(2.13); and an inversion of the latter discloses that

$$r_{1,2}^2 = \frac{C_{1,2}^2}{(1 - C_3)C_1 + C_2^2} \tag{3.2}$$

and

$$\sin^2 j = \frac{C_1}{(1 - C_3)C_1 + C_2^2}. \tag{3.3}$$

Since $\sin^2 j \leq 1$, this last equation discloses that $C_2^2 \leq C_1 C_3$; for central eclipses (when the equality sign holds) C_2 thus represents the geometrical mean of C_1 and C_2. Lastly, for total eclipses (when $r_2 - r_1 > \cos j$), all three C_j's are bound to be positive, and such that $C_1 < C_2 < C_3$. Should $\cos j = r_2 - r_1$ (grazing eclipse), $C_3 = C_1 + 2C_2$; and if $r_1 = r_2$ (i.e., $k = 1$)—a situation approximated by total eclipses of the Sun by our Moon—$C_1 = C_2$ while $C_3 = 3C_2$.

The question can be asked: is it possible to extend this solution by including the degree(s) of limb-darkening u_n among the unknowns to be determined by simultaneous analysis? The answer is indeed affirmative in principle, but not so easy to carry out in practice; and to do so requires a knowledge of the moments A_{2m} of the respective light curve beyond those corresponding to $m = 1, 2$ and 3. In fact, should we choose to describe the distribution of surface brightness J over the apparent disc of the star undergoing eclipse by N distinct coefficients u_n on the right-hand side of Eq. (1.2) of Chapter II, and treat them as unknowns to be determined simultaneously with three geometrical elements $r_{1,2}$ and i of the eclipse, a total of $3 + N$ unknowns would call for an empirical knowledge of not less than $3 + N$ moments A_{2m} ($m = 1, 2, 3, \ldots N + 3$).

The theoretical values of the requisite higher moments A_{2m} can be determined in the same way followed in the preceding section, which led to (1.25)–(1.27). In particular, for $N = 1$ and $m = 4$ we find that

$$A_8 = L_1 \left\{ \overline{C}_3^4 + 4\overline{C}_1 \overline{C}_2^2 \overline{C}_3 + 6\overline{C}_2^2 \overline{C}_3^2 + \right.$$

$$+ \frac{30}{7} \frac{(3 - u_1)(35 - 19u_1)}{(15 - 7u_1)^2} \overline{C}_2^4 +$$

$$+ \frac{7}{27} \frac{(15 - 7u_1)(315 - 187u_1)}{(35 - 19u_1)^2} \overline{C}_1^2 \overline{C}_2^2 \bigg\} ; \tag{3.4}$$

and with the values of $\overline{C}_{1,2,3}$ already known with Eqs. (1.25)–(1.27), the foregoing Equation (3.4) can be solved for u_1 as the only unknown; though observational data of very high precision would be needed to do so significantly.

If, however, the eclipses in question are *partial* (or annular), the situation becomes more complicated. This complication enters at the beginning of our procedure; for if $\delta_0 > | r_1 - r_2 |$, Equation (1.5) reduces to

$$1 - \lambda = A_0 , \tag{3.5}$$

where A_0 is given by Equation (2.3) in which we have set $m = 0$. Unlike in the case of total eclipses this, however, will be no longer a constant directly deducible from the observations, but remains—like all other moments A_{2m} for $m > 0$—also functions of the geometrical elements $r_{1,2}$ and j of the eclipse, to be determined simultaneously with all others.

In the case of total eclipses, the empirical value of A_0 is, in principle, sufficient to determine L_1, while the moments corresponding to $m = 1, 2$ and 3 then specify $r_1,, r_2$ and j. However, for eclipses other than total, the empirical values of *four* A_{2m}'s are prerequisite for setting up a simultaneous system of four equations to specify $r_{1,2}$, j *and* L_1 at the same time. Moreover—and unlike in the case of total eclipses—the solution for the elements of partially-eclipsing systems cannot be carried out in a closed form; and the only feasible approach towards it is by *iteration*. Moreover, in order to render such a solution determinate, consideration should be given to an obvious possibility to combine photometric evidence bearing on the *alternate* minima, in which the role of the fractional radii $r_{1,2}$ of the components of the system, and their fractional luminosities $L_{1,2}$ become *interchanged*.

In more specific terms, let us express Equation (2.3) in the form

$$(A_{2m})_{\text{pri}} = L_1 b^2 \sin^{2m} \psi_1 f_{2m}(b, c_0) \tag{3.6}$$

at the time of the primary (deeper) minimum; while half a revolution later (when the star of radius r_1 and luminosity L_1 is now in front)

$$(A_{2m})_{\text{sec}} = L_2 a^2 \sin^{2m} \psi_1 f_{2m}(a, c_0). \tag{3.7}$$

One way to proceed with their solution is to eliminate the fractional luminosities $L_{1,2}$ between them with the aid of Eq. (1.3). The linear ratio of two moments

$$\frac{A_{2m}}{A_{2\mu}} \equiv \sin^{2(m-\mu)} \psi_1 \frac{f_{2m}(b, c_0)}{f_{2\mu}(b, c_0)} \tag{3.8}$$

depend only on ψ_1, $b\,(a = 1 - b)$ and c_0. However, the quadratic ratios

$$\frac{A_{2m}A_{2m'}}{A_{2\mu}A_{2\mu'}} \equiv \sin^{2(m+m'-\mu-\mu')}\psi_1 \times$$

$$\times \frac{f_{2m}(b,c_0)f_{2m'}(b,c_0)}{f_{2\mu}(b,c_0)f_{2\mu'}(b,c_0)}, \tag{3.9}$$

where m, m' and μ, μ' are all distinct, but such that

$$m + m' = \mu + \mu', \tag{3.10}$$

the right-hand side of Eq. (3.8) depends only on b and c_0.

Suppose that, in particular, $m = m' = 1$ but $\mu = 2$ and $\mu' = 0$. If so,

$$\frac{(A_2)^2}{A_0 A_4} \equiv \frac{A_2/A_0}{A_4/A_2} \equiv g_2(b,c_0) \tag{3.11}$$

while if $m = m' = 2$ and $\mu = 3$, $\mu' = 1$

$$\frac{(A_4)^2}{A_2 A_6} \equiv \frac{A_4/A_2}{A_6/A_4} \equiv g_4(b,c_0). \tag{3.12}$$

With the left-hand sides of the preceding equations ascertained from the observations, the functions $g_{2,4}\,(b,c_0)$ constitute two independent relations between the unknown parameters b and c_0; and can be solved for them (numerically or otherwise) with the aid of the expansions on the r.h. sides of Eq. (2.3); the necessary and sufficient condition for such a solution being the requirement that the Jacobian determinant

$$\frac{\partial(g_2, g_4)}{\partial(b, c_0)} \neq 0 \tag{3.13}$$

i.e., that the functions $g_{2,4}$ do not simulate too closely functional dependence. It is, in fact, this condition which should enable us to *optimize* our solution by *maximizing* the Jacobian (3.13)—a task calling for a judicious choice of the constants m and μ (which need not necessarily be integers); and facilitated by the excellent *asymptotic properties* of the expansion on the r.h.s. of (2.3), for the details of which cf. Kopal (1986).

It goes without saying that, in general, the construction of such a solution can be approached only by *iteration*. It is, however, not necessary for this purpose to eliminate the fractional luminosities $L_{1,2}$ from Eqs. (3.6) and (3.7) to begin with, to iterate the solution of (3.9). For although the value of L_1 may be difficult to estimate in advance, the phase-angle ψ_1 of the first contact of the eclipse is not. Estimating it, we can commence our iterative process for ψ_1, b and c_0 from Equation (3.8), and once this has been accomplished for a judicious choice of m and μ, the geometrical elements r_1, r_2 and j then follow from Eqs. (2.4) as

$$r_1^2 = \frac{a^2 \sin^2 \psi_1}{1 - c_0^2 \cos^2 \psi_1}, \tag{3.14}$$

$$r_2^2 = \frac{b^2 \sin^2 \psi_1}{1 - c_0^2 \cos^2 \psi_1} \tag{3.15}$$

and

$$\sin^2 j = \frac{1 - c_0^2}{1 - c_0^2 \cos^2 \psi_1} \tag{3.16}$$

($a \equiv 1-b$)—replacing Eqs. (3.2)–(3.3), in which the same elements were expressed in terms of the auxiliary constants $C_{1,2,3}$—and

$$\sin^2 \psi_1 = (r_1 + r_2)^2 \csc^2 j - \cot^2 j \equiv C_1 + 2C_2 + C_3. \tag{3.17}$$

The values of r_1, r_2 and $\sin j$ resulting from the above equations (3.14)–(3.16) should than be inserted in the right-hand side of (3.17), and the value of $\sin^2 \psi_1$ so obtained compared with the one adopted at the outset. Should these disagree to any significant extent, the new value should replace the one previously adopted; and the process repeated until both the incoming and outgoing values have been brought to agree.

The sucess of this process presupposes, of course, that a selected pair of m and μ used to set up equations of the form (3.8) leads to a satisfactory solution for b and c_0. This can, in general, be relied upon only for totally-eclipsing systems. If the eclipses are partial it has been known since the time of Russell and Shapley in 1912 that (unless the underlying observations are accurate to better than 1 part in 1000) no reliable solution for the elements of the system can be extracted from the evidence furnished by one minimum alone; and that, in order to arrive at it, at least some knowledge of the alternate (secondary) minimum becomes a necessary prerequisite.

If so, therefore, it becomes necessary to resort to another set of relations of the form

$$\frac{(A_{2m})_{\text{pri}}}{(A_{2m})_{\text{sec}}} = \frac{L_1}{L_2} \left(\frac{r_2}{r_1}\right)^2 \frac{f_{2m}(b, c_0)}{f_{2m}(a, c_0)}, \tag{3.18}$$

which should be adjoined to Eq. (3.8) to increase the determinacy of our solution. The right-hand side of the foregoing equation still contains the ratio L_1/L_2. Its value can, however, be readily ascertained if we particularize Eq. (3.18) for $m = 0$, when $f_0(a \text{ or } b, c_0)$ stands for the maximum obscuration at the time of the respective eclipse. Since, moreover

$$(A_0)_{p,s} = 1 - \lambda_{p,s}, \tag{3.19}$$

a combination of (3.17) and (3.18) discloses that

$$\frac{L_1}{L_2} = \left(\frac{r_1}{r_2}\right)^2 \left\{\frac{f_0(a, c_0)}{f_0(b, c_0)}\right\} \frac{1 - \lambda_{\text{pri}}}{1 - \lambda_{\text{sec}}} \equiv$$

$$\equiv \left(\frac{r_1}{r_2}\right)^2 Y(a, c_0) \frac{1 - \lambda_{\text{pri}}}{1 - \lambda_{\text{sec}}}, \tag{3.20}$$

where $Y(a, c_0)$ stands for a slowly-varying function of its parameters.

Next, let us return to Eqs. (3.6)–(3.7), and particularize these for $m = 0$: since

$$(A_0)_{\text{pri}} = 1 - \lambda_p = L_1 b^2 f_0(b, c_0) \qquad (3.21)$$

and

$$(A_0)_{\text{sec}} = 1 - \lambda_s = L_2 a^2 f_0(a, c_0) \qquad (3.22)$$

their combination with Eqs. (1.3) and (3.20), (3.21) discloses that

$$\alpha_0 \equiv b^2 f_0(b, c_0) = 1 - \lambda_p + \frac{1 - \lambda_s}{k^2 Y(a, c_0)} , \qquad (3.23)$$

where α_0 stands for the maximum obscuration of the star of fractional radius r_1, and (as before) $k \equiv r_1/r_2$. Moreover, for any value of $m > 0$, Equation (3.18) on insertion for L_1/L_2 from (3.20) yields

$$\frac{(A_{2m})_p (A_0)_s}{(A_{2m})_s (A_0)_p} = \frac{f_0(b, c_0) f_{2m}(a, c_0)}{f_0(a, c_0) f_{2m}(b, c_0)} , \qquad (3.24)$$

the left-hand side of which can be determined from the light curve if (at least) the depths of both minima are known. Should the latter become negligible (as it would be if the components eclipsed at that time were effectively dark and, therefore, $\lambda_s = 1$),

$$\alpha_0(k, c_0) \equiv b^2 f_0(b, c_0) = 1 - \lambda_p , \qquad (3.25)$$

which together with Eqs. (3.11) or (3.12) can then specify the values of $r_{1,2}$ and j.

V.4 Moments of the Light Curves: Determination from the Observations

In the preceding section of this chapter we have the way in which the moments of the light curves A_{2m} as defined by Equation (1.10) can be utilized to specify the values of the fractional radii $r_{1,2}$ and of the inclination j of their orbit. By this our task has not come yet to the end; for it still remains for us to specify the *uncertainty* within which the moments A_{2m} of the observed light changes are specified by the available observational data. The extent of this uncertainly is, of course, implied in the photometric observations at the basis of our study—in their quality as well as numbers—but appropriate procedures towards this end remain still to be described.

Since astronomical observations (or, indeed, physical measurements of any kind) are never infinitely numerous or precise, all results based upon them are bound to be inaccurate within finite limits; and their uncertainty is usually described by their mean (or probable) errors in terms of the theory of "least squares". Such errors—it cannot be emphasized too often—should be regarded as an inseparable part of each solution; and their specification becomes, in fact, the more important, the fewer (or less accurate) observations are at our disposal; for the danger of overestimating the significance of the results based upon them becomes then greatest.

In our present problem, the sources of this uncertainty can be split up in two parts: (a) a determination of the uncertainty of the empirical moments A_{2m} of the light curves as defined by available observational data; and (b) a translation of this uncertainty into those of the geometrical elements $r_{1,2}$ and i of the respective system. To consider the source (a) first, suppose that the variation of light $l(\psi)$ with the phase ψ during minima can be expressed by a Fourier series of the form

$$1 - l = \frac{1}{2}a_0 + \sum_{n=1}^{N} a_n \cos(n\pi\psi/\psi_1) , \tag{4.1}$$

where ψ_1 stands, as before, for the angle of first contact of the eclipse, and the a_n's are the respective Fourier coefficients. As many equations of condition (say, M) of the form (4.1) can obviously be set up as there are observed values of $l(\psi)$, and solved by the method of least squares for the most probable values of $N + 1$ unknown constants a_n; together (if $M \gg N$) with the uncertainty within which they are defined by the available observational data. And once we have done so, the moments A_{2m} as defined by Eq. (1.1) can be expressed in their terms by the equation

$$\begin{aligned}
A_{2m} &= \int_0^{\psi_1} (1 - l)d(\sin^{2m}\psi) = \\
&= \frac{1}{2}a_0 \sin^{2m}\psi_1 + \\
&\quad + \sum_{n=1}^{N} a_n \int_0^{\psi_1} \cos(n\pi\psi/\psi_1)d(\sin^{2m}\psi) ,
\end{aligned} \tag{4.2}$$

where the constant $m > 0$ stands for an arbitrary (non-negative) number.

In what follows let us, however, restrict the admissible values of m to non-negative integers. If so, the integrals on the right-hand side of the preceding equation can be evaluated in a closed form to yield

$$A_0 = \sum_{n=0}^{N} \frac{\epsilon_n}{2}a_n , \tag{4.3}$$

$$A_2 = \sum_{n=0}^{N} \left\{ \frac{\epsilon_n}{2} \frac{a_{2n}\sin^2\psi_1}{1^2 - [n\pi/\psi_1]^2} + \frac{a_{2n+1}\cos^2\psi_1}{1^2 - [(2n+1)\pi/2\psi_1]^2} \right\} , \tag{4.4}$$

$$A_4 = \sum_{n=0}^{N} \frac{\epsilon_n}{2} \left\{ \frac{\sin^2\psi_1}{1^2 - [n\pi/\psi_1]^2} + \frac{\sin^2 2\psi_1}{2^2 - [n\pi/\psi_1]^2} \right\} a_{2n} +$$

$$+ \sum_{n=0}^{N} \left\{ \frac{\cos^2 \psi_1}{1^2 - [(2n+1)\pi/2\psi_1]^2} - \frac{\cos^2 2\psi_1}{2^2 - [(2n+1)\pi/2\psi_1]^2} \right\} a_{2n+1}, \quad (4.5)$$

$$A_6 = \frac{3}{16} \sum_{n=0}^{N} \frac{\epsilon_n}{2} \left\{ \frac{5 \sin^2 \psi_1}{1^2 - [n\pi/\psi_1]^2} - \frac{8 \sin^2 2\psi_1}{2^2 - [n\pi/\psi_1]^2} + \frac{3 \sin^2 3\psi_1}{3^2 - [n\pi/\psi_1]^2} \right\} a_{2n} +$$

$$+ \frac{3}{16} \sum_{n=0}^{N} \left\{ \frac{5 \cos^2 \psi_1}{1^2 - [(2n+1)\pi/2\psi_1]^2} - \frac{8 \cos^2 2\psi_1}{2^2 - [(2n+1)\pi/2\psi_1]^2} + \right.$$

$$\left. + \frac{3 \cos^2 3\psi_1}{3^2 - [(2n+1)\pi/2\psi_1]^2} \right\} a_{2n+1}, \quad (4.6)$$

etc., where $\epsilon_n = 1$ for $n = 0$ and 2 for $n > 0$.

Equations (4.3)–(4.6) express the moments A_{2m} of the light curves in the form of linear functions of the coefficients A_n; and their uncertainty can be specified of that of the respective linear functions—with due regard to the fact that the errors of the A_n's (resulting as they do from a simultaneous solution of the equations of condition of the form (4.1) are not independent. The same is, of course, also true of the uncertainty of the elements $r_{1,2}$ and j—or, rather, of the linear functions $\delta r_{1,2}$ and δj obtained by differentiation of the expressions of the respective elements. Explicit expressions for $\delta r_{1,2}$ and δj have already been recently described elsewhere (cf. sec. 5 of Kopal, 1986) and need not be repeated in this place.

All such results obtained so far are, of course, dependent on the value of the angle ψ_1 on the right-hand sides of Eqs. (3.27)–(3.31) originally estimated from the light curve. Once a preliminary set of the elements has thus been obtained, their insertion in Eq. (3.17) can verify the extent to which our original estimate was correct. Should the difference between the incoming and outgoing value of ψ_1 be significant, the process should be repeated until the two no longer differ to a significant extent.

It is of course possible to by-pass the complication arising from the presence of ψ_1 on the right-hand sides of the expressions (3.28)–(3.31) by replacing the use of these formulae for the A_{2m}'s by *graphical* integration as represented by *planimetry*. To this end it should be sufficient to plot the observed light intensity $l(\psi)$ against $\sin^{2m} \psi$ (see Figure V.1), and draw a smooth curve through the individual observations to specify the area between the lines $l = 1$, $\sin^{2m} \psi = 0$, and the actual light curve within minima; to ascertain the area subtended by these limits by a process no more elaborate than the "counting of the squares" on the millimetre-paper of such a plot. For such a process (tantamount to a "box-car" integration with an arbitrarily small interval) can, in competent hands, furnish likewise respectable results.

However—and this is essential—mere planimetry of this type can never disclose the true uncertainty of the moments A_{2m} reflecting the full extent of observational errors. It is true that the $\sin \psi$-coordinates—depending only on the timing of the observations—are the subject of errors which (for all practical purposes) can be regarded as negligible; and the maximum brightness $l = 1$ of the system (for those consisting of spherical stars) can be ascertained as the (weighted) mean of all observations made outside eclipses; the uncertainty is relatively easy to establish. But the branch of the area delimited by the ascending or descending branch of the light curve—affected as it is by a dispersion of the individual observations defining it—is more difficult to take into account; for a more quantitative analysis the numerical procedures described earlier in this section become indispensable.

V.5   Bibliographical Notes

The introduction of the moments of the light curves A_{2m} as defined by Eq. (1.1)—tantamount to a transfer of our problem from the time- to the frequency-domain (the index m playing the role of a frequency)—goes back to 1975 (cf. Kopal, 1975a,b) where a proof has been given that, for total eclipses, the relations between the A_{2m}'s and the elements of the system L_1, $r_{1,2}$ and of the orbital inclination $n_0 \equiv \cos j$ can be formulated algebraically in a closed form; subsequently, the process has been extended to any type of eclipse (and to an arbitrary degree of darkening of the star undergoing eclipse) in Kopal (1977b).

For a general theory of functions satisfying the difference-differential equation (2.24), of which the moments A_{2m} represent a particular case, cf. Appell (1880), Humbert (1924), Milne-Thomson (1933) or Truesdell (1948); the same equation is satisfied (for the same reasons) also by the photometric perturbations $\mathcal{P}_{2m}$, defined by Equation (2.4) of Chapter VI.

For additional types of recursion formulae satisfied by the moments A_{2m} cf. Appendix II of Kopal (1982d).

The algebraic processes by which these moments can be inverted to furnish the geometrical elements of the eclipses cf. Kopal and Demircan (1978) and subsequent papers, among which those by Kopal (1982a and 1986) are the latest to be concerned with mainly practical aspects of the subject. For a detailed derivation of Eqs. (3.28)–(3.31) see Kopal, 1982b).

Chapter VI

INVERSE PROBLEM FOR DISTORTED ECLIPSING SYSTEMS

In the preceding Chapters II-III and V of this book we have outlined the basic principles of analysis of the light curves of eclipsing binary systems, appropriate for the case in which their components can be regarded as spheres. To what extent can, however, such a model be regarded as a satisfactory representation of eclipsing systems actually observed in the sky? In many of them, the components are indeed separated widely enough—and, as a result, the photometric effects caused by their departures from spherical form are small enough for the methods already expounded to meet all our needs. For observations accurate to about one part in 1000; this should in general be true as long as their fractional radii $r_{1,2}$ of the components constituting such systems do not exceed (approximately) 0.15 in numerical value.

However, if the fractional dimensions of the components exceed markedly this limit, the photometric proximity effects are no longer ignorable; and, as a result, the methods of solution expounded in the preceding chapters of this book require generalization. As is well known, a tell-tale feature of such "close" eclipsing systems is the fact that their light changes are no longer confined only to the times of the minima, but extend over the whole cycle. Moreover—in contrast with photometric effects of ellipticity and reflection which vanish only if the orbital plane is perpendicular to the line of sight (and not even then if the relative orbit of the two stars is eccentric), those due to eclipses are discontinuous, and can last only a fraction of each cycle. In moderately close systems (characterized by fractional radii between 0.25–0.30), it is possible to discern by an inspection of their light curves when the eclipses set in; and an example of such a case is shown on the accompanying Figure VI.1. However, in really close systems (with fractional radii of the components exceeding 0.3), the amplitudes of the light changes arising from both these sources may become comparable—making it impossible to decide by a mere inspection of their light curves when (or whether) any eclipses actually set in; or whether the observed light variations are due to the proximity effects alone (for an example of such a case see Figure VI.2). Such systems are, moreover, very common—both in the sky and in the variable star catalogues (because the proximity of their components favours discovery); and their interpretation confronts us, therefore, with an important problem.

It is at this stage that Nature lends us once more a very helpful hand: for *even if the components of close binary systems are in actual contact, their light*

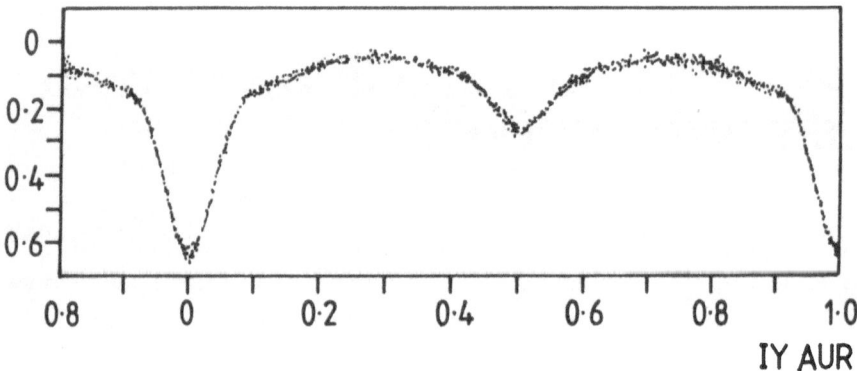

Figure VI.1: Light changes exhibited by IY Aurigae—a moderately close eclipsing binary system—and reproduced from Fracastoro's *Atlas of the Light Curves of Eclipsing Binaries* (Torino, 1972) based on the observations by P. Tempesti.

changes cannot be affected by any eclipses within a definite phase range on either side of the quadratures, regardless of the degree of proximity of the components of the respective system. The reason for this fortunate circumstance (first pointed out by the present writer in 1954) goes back to the geometry of the Roche model (representing the shape of centrally-condensed stars to a high degree of accuracy) which discloses that, for a very wide range of the mass-ratios (exceeding 10:1) the light changes of contact (let alone detached) systems are unaffected by eclipses at all phases in excess of ±58° around each conjunction even if their orbital planes were perpendicular to the celestial sphere. Therefore, such light changes as are exhibited between phase angles 58° and 122 ° around the first quadrature, and within the same interval half a revolution later, must be due solely to the proximity effects, and without any interference from eclipses.

For the latest discussion of this geometry cf. Chapter II of Kopal (1989), to which the reader is referred for fuller details. The facts summarized there would seem to make it possible to separate in time the photometric manifestations of the proximity effects (extending over the entire cycle, but confined in pure form to the neighbourhood of quadratures) from the eclipse effects (which are bound to remain restricted to the neighbourhood of conjunctions); but the methods by which this can be done require careful consideration.

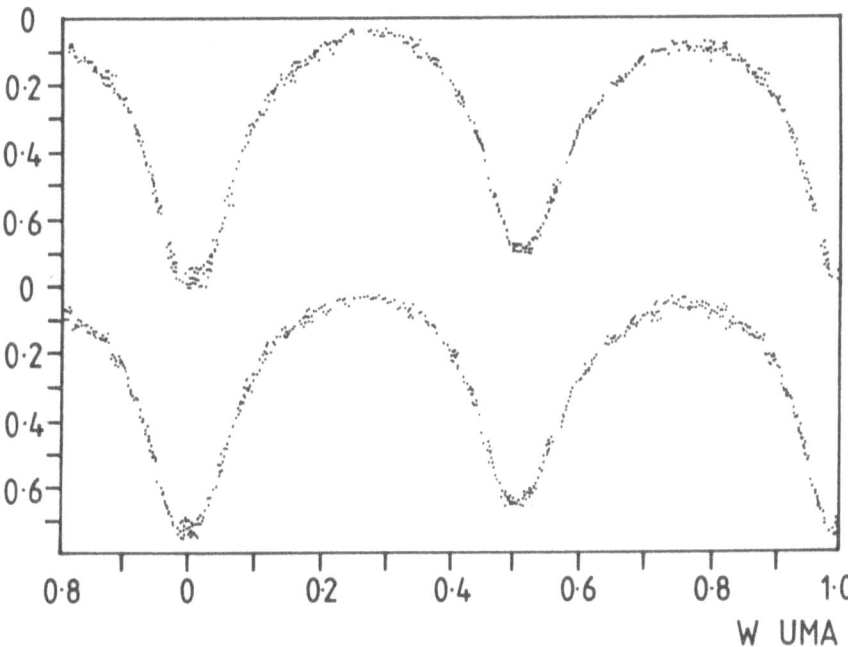

Figure VI.2: Light changes exhibited by the close eclipsing system W Ursae Maioris (after L. Binnendijk, *Astron. J.*, **71**, 340, 1960).

VI.1   Variation of Light Between Eclipses

In order to do so, let us confine first our attention to the variation of light $\mathcal{L}(l_0)$ *between minima* to be given by

$$\mathcal{L} \equiv \mathcal{L}_1(l_0) + \mathcal{L}_2(-l_0) \,, \tag{1.1}$$

where (to the first order in small quantities), the theoretical values of $\mathcal{L}_{1,2}(\pm l_0)$ is given by Eq. (2.46), Chapter VII of Kopal (1989); with due regard being paid to the fact that the phases of the two components differ by 180°, and their respective direction cosines of l_0 are of opposite sign. Therefore, the equation (2.46) referred to above can be theoretically rewritten as

$$\mathcal{L} = \sum_{j=0}^{n} c_j \cos^j \psi \tag{1.2}$$

for the range in phase ψ independent of eclipses, and the C_j's are constants deducible from (2.46).

Within that range, as many equations of the form (1.2) can obviously be set up as there are observed values of $l(\psi)$ on hand, and solved for the coefficients c_j by the method of least squares. In doing so we run, however, into a difficulty unforeseen by the earlier investigators: namely, that *with increasing value of n, the determinant of the system of normal equations from which the c_j's are to be obtained will rapidly approach zero*—a circumstance tending to render their solution indeterminate. The cause of this unpleasant phenomenon is the fact that the powers of $\cos \psi$ constitute an orthogonal set of functions which are independent over a closed interval (-1, 1), but their orthogonality fails when this interval is restricted to an eclipse-free zone; while over a more limited interval they begin to *simulate functional dependence* that brings about a disastrous loss of weight of the entire solution.

In order to demonstrate the extent to which this is true, consider a family of Tchebyshev polynomials $T_j(x)$ of the first kind, defined by

$$T_j(x) = \cos jt, \quad \cos t = (2x - b - a)/(b - a) \tag{1.3}$$

where $x \equiv \cos \psi$ and (a, b) represent the limits of their orthogonality. Earlier in this section we mentioned already that, in close binary systems exhibiting continuous light variation throughout the cycle, it may not be safe to assume the limits in phase free from eclipse effect to go beyond $90° \pm 32°$, or (to remain on the safe side) to $90° \pm 30°$ rendering

$$(a, b) = \cos(90° \pm 30°) = \pm\frac{1}{2}. \tag{1.4}$$

Accordingly, Eq. (1.3) then reduces to

$$T_j(\cos \psi) = \cos j \cos^{-1}(2 \cos \psi), \tag{1.5}$$

which for $j = 4, 5, 6$ assume the explicit forms:

$$T_4(\cos \psi) = 128 \cos^4 \psi - 32 \cos^2 \psi + 1, \tag{1.6}$$

$$T_5(\cos \psi) = 512 \cos^5 \psi - 160 \cos^3 \psi + 10 \cos \psi, \tag{1.7}$$

$$T_6(\cos \psi) = 2048 \cos^6 \psi - 768 \cos^4 \psi + 72 \cos^2 \psi - 1, \tag{1.8}$$

etc.

Since, however, it follows from Eq. (1.2) that

$$| T_j(\cos \psi) | \leq 1 \tag{1.9}$$

for any argument $\cos \psi$ within the limits of $\pm\frac{1}{2}$—and, therefore, the left-hand sides of Equations (1.6)–(1.9)—the latter can be solved for the highest powers of cos ψ involved in them to yield

$$\cos^4 \psi = \frac{1}{4} \cos^2 \psi - \frac{1}{128} , \tag{1.10}$$

$$\cos^5 \psi = \frac{5}{16} \cos^3 \psi - \frac{5}{256} \cos \psi , \tag{1.11}$$

$$\cos^6 \psi = \frac{3}{8} \cos^4 \psi - \frac{9}{256} \cos^2 \psi + \frac{1}{2048} ; \tag{1.12}$$

which within the range $60° \leq \psi \leq 120°$ are subject to errors not exceeding 2^{-7}, 2^{-9} and 2^{-11}, respectively. Outside this range, the foregoing approximations (1.10)–(1.12) deteriorate and eventually become useless; but within the range represented by (1.4), higher powers of cos ψ are seen to simulate linear combinations of lower powers with remarkable fidelity; and *this* causes the systems of normal equations based on (1.2) to become so ill-conditioned with increasing value of n.

It may, of course, happen that the eclipse-free portion of the light curve extends beyond the interval from 60° to 120° in phase—as it can happen if one (or both) components are of the detached type; or the orbital inclination is appreciably less than 90°; or both. In such a case, the limits of orthogonality in (1.2) can be extended with impunity without running into the eclipse effects; and the quality of the solution for the c_j's increased.

For example, if the eclipse-free zone extends between

$$45° < \psi < 135° \tag{1.13}$$

corresponding to the orthogonality limits

$$(a, b) = \pm \frac{\sqrt{2}}{2} , \tag{1.14}$$

Tchebyshev polynomials for $j = 4, 5, 6$ become

$$T_4(\cos \psi) = \cos 4 \sin^{-1} \sqrt{2} \cos \psi = {}_2F_1(-2, 2; \frac{1}{2}; 2 \cos^2 \psi), \tag{1.15}$$

$$T_5(\cos \psi) = \sin 5 \sin^{-1} \sqrt{2} \cos \psi =$$
$$= 5\sqrt{2} \cos \psi \, {}_2F_1(-2, 3; \frac{3}{2}; 2 \cos^2 \psi), \tag{1.16}$$

$$T_6(\cos \psi) = -\cos 6 \sin^{-1} \sqrt{2} \cos \psi = -{}_2F_1(-3, 3; \frac{1}{2}; 2 \cos^2 \psi); \tag{1.17}$$

where ${}_2F_1$ stands for a (terminating) hypergeometric series (i.e., Jacobi polynomials).

Should, moreover, the eclipse-free portion of the light curve extend to

$$30° < \psi < 150° \tag{1.18}$$

corresponding to the orthogonality limits

$$(a, b) = \pm\frac{\sqrt{3}}{2}, \tag{1.19}$$

the Tchebyshev polynomials $T_j(\cos \psi)$ appropriate for $j = 4, 5, 6, \ldots$ assume the forms

$$T_4(\cos \psi) = \cos 4 \sin^{-1} \frac{2}{\sqrt{3}} \cos \psi = {}_2F_1(-2, 2; \frac{1}{2}; \frac{4}{3} \cos^2 \psi), \tag{1.20}$$

$$T_5(\cos \psi) = \sin 5 \sin^{-1} \frac{2}{\sqrt{3}} \cos \psi =$$

$$= \frac{10}{\sqrt{3}} \cos \psi \, {}_2F_1(-2, 3; \frac{3}{2}; \frac{4}{3} \cos^2 \psi), \tag{1.21}$$

$$T_6(\cos \psi) = -\cos 6 \sin^{-1} \frac{2}{\sqrt{3}} \cos \psi = -{}_2F_1(-3, 3; \frac{1}{2}; \frac{4}{3} \cos^2 \psi); \tag{1.22}$$

yielding the approximations

$$\cos^4 \psi = \frac{1}{2} \cos^2 \psi - \frac{1}{32} \pm 2^{-5} \, (45° < \psi < 135°)$$

$$= \frac{3}{4} \cos^2 \psi - \frac{9}{128} \pm 9 \times 10^{-7} \, (30° < \psi < 150°) \tag{1.23}$$

$$\cos^5 \psi = \frac{5}{8} \cos^3 \psi - \frac{5}{64} \cos \psi \pm \frac{2^{-6}}{\sqrt{2}} \, (45° < \psi < 135°)$$

$$= \frac{15}{16} \cos^3 \psi - \frac{45}{256} \cos \psi \pm 9\sqrt{3} \times 2^{-9} \, (30° < \psi < 140°) \tag{1.24}$$

and

$$\cos^6 \psi = \frac{3}{4} \cos^4 \psi - \frac{9}{64} \cos^2 \psi + \frac{1}{256} \pm 2^{-8} \, (45° < \psi < 135°) \tag{1.25}$$

$$+ \frac{9}{8} \cos^4 \psi - \frac{81}{256} \cos^2 \psi + \frac{27}{2048} \pm 27 \times 2^{-11} \, (30° < \psi < 150°).$$

The foregoing results make it clear that to determine significant values of the constants c_j on the right-hand side of Eq. (1.2) from the observed values of $\mathcal{L}(\psi)$ between eclipses will not be easy—in fact, seldom more than three such constants (for $n = 2$) can be so determined. In order to *force* so truncated a polynomial of the form

$$\mathcal{L} = C_0 + C_1 \cos \psi + C_2 \cos^2 \psi, \tag{1.26}$$

through the observations, and solve for the empirical values of $C_{1,2,3}$ by the method of least-squares, an insertion of (say) Eqs. (1.10)–(1.12) for higher powers of $\cos^j \psi$ in (1.2) discloses that the coefficients C_j of so "telescoped" a series of the form (1.26) become

$$C_0 \;=\; c_0 - \frac{1}{128}c_4 + \frac{1}{2048}c_6 + \cdots, \tag{1.27}$$

$$C_1 \;=\; c_1 \qquad\qquad - \frac{5}{256}c_5 + \cdots, \tag{1.28}$$

$$C_2 \;=\; c_2 + \frac{1}{4}c_4 - \frac{9}{256}c_6 + \cdots, \tag{1.29}$$

$$C_3 \;=\; c_3 + \frac{5}{16}c_5 \cdots, \tag{1.30}$$

$$C_4 \;=\; c_4 + \frac{3}{8}c_6 \cdots; \tag{1.31}$$

and if, consistent with (1.26) we have set

$$C_3 \;=\; C_4 \;=\; 0, \tag{1.32}$$

it would follow that

$$C_0 \;=\; c_0 - \frac{7}{768}c_4 + \cdots, \tag{1.33}$$

$$C_1 \;=\; c_1 + \frac{1}{16}c_3 + \cdots, \tag{1.34}$$

$$C_2 \;=\; c_2 + \frac{11}{32}c_4 + \cdots; \tag{1.35}$$

and similar expressions can be obtained by a resort to (1.23)–(1.25) in place of (1.10)–(1.12)—depending on the segment of the light curve believed to be unaffected by eclipses. As can be seen, a physical interpretation of the constants $C_{0,1,2}$ is by no means straightforward; and care should be exercised in carrying out such an operation.

VI.2 Modulation of the Light Curves

Whatever a knowledge of the constant c_j on the r.h.s. of Eq. (1.2) can tell us about the physical properties of the respective system, it can only indirectly serve to a determination of the geometrical system. These can only come out of the empirically-determined moments of the eclipses (if any), representing the areas subtended by the light changes in the $\mathcal{L}(\psi) - \sin^{2m}\psi$ plane. For systems consisting of spherical stars (see Sec. V-1), a determination of these moments has already been discussed in the preceding chapter of this book. In systems whose components are appreciably distorted by rotational and tidal forces, however, the limiting line $l = 1$ is to be replaced by a curve $\mathcal{L}(\psi)$ as given by Eq. (1.1); the coefficients of which have been determined (numerically or otherwise) from the light changes exhibited by such systems between minima.

If an extrapolation of Eq. (1.1) determined (say) within the limits $60° < \psi < 120°$—a safe process even if the components are in actual contact—can represent the light changes also for $0° < \psi < 60°$ without significant residuals, this means that no eclipses occur and the observed light changes are due to the proximity effects alone. Only if this is not the case, eclipses set in and Equation (1.2) of Chapter V is then to be augmented to

$$\Delta\mathcal{L} \equiv \mathcal{L}\left(\frac{\pi}{2}\right) - \mathcal{L}(\psi) =$$

$$= \sum_{j=0}^{n} c_j \cos^j \psi + L_1 \sum_{n=0}^{N} C^{(n)} \alpha_n^0 +$$

$$+ L_1 \sum_{n=0}^{N} C^{(n)} \{f_1^{(n)} + f_2^{(n)}\}, \qquad (2.1)$$

in which the $f_{1,2}^{(n)}$'s stand for the "photometric perturbations during eclipses", arising from the distortion of the two components, as given by Equations (2.21) and (3.31) of Chapter IV.

Let us, moreover, extend now the definition (1.1) in Chapter V of the empirical moments $\overline{A}_{2m}$ of the light curves of distorted eclipsing systems (cf. Fig. VI.3) to

$$\overline{A}_{2m} \equiv \int_0^{\pi/2} \left\{ \mathcal{L}\left(\frac{\pi}{2}\right) - \mathcal{L}(\psi) \right\} d \sin^{2m} \psi, \qquad (2.2)$$

which on insertion from (2.1) can be rewritten as

$$\overline{A}_{2m} = \sum_{j=0}^{n} c_j \int_0^{\pi/2} \cos^j \psi \, d(\sin^{2m} \psi) +$$

$$+ L_1 \sum_{n=0}^{N} C^{(n)} \int_0^{\pi/2} \left\{ \alpha_n^0 + f_1^{(n)} + f_2^{(n)} \right\} d(\sin^{2m} \psi) =$$

$$= m \sum_{j=1}^{N} B\left(\frac{j+2}{2}, m\right) c_j + A_{2m} + \mathcal{P}_{2m}, \qquad (2.3)$$

where A_{2m} continues to be defined by Eq. (1.1), and where we have abbreviated

$$\mathcal{P}_{2m} = L_1 \sum_{n=0}^{N} C^{(n)} \int_0^{\psi_1} \{f_1^{(n)} + f_2^{(n)}\} d(\sin^{2m} \psi), \qquad (2.4)$$

in which (since $f^{(n)}(\psi) \equiv 0$ for $\psi \geq \psi_1$) the upper limit of integration can be restricted from 90° to the phase angle ψ_1 of first contact of the eclipse.

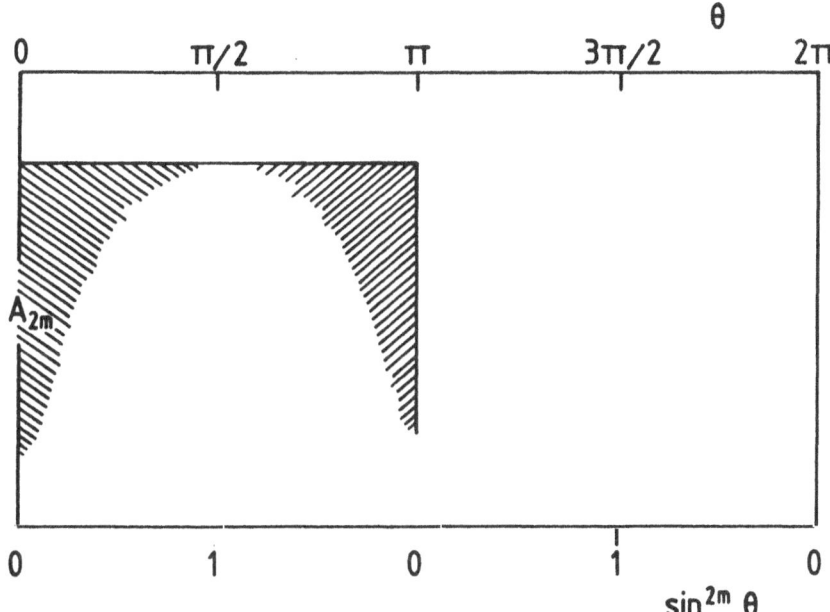

Figure VI.3: Areas representing the moments A_{2m} of the light curve of the eclipsing system W Ursae Maioris, shown on Figure VI.2. Separate moments can be determined for the primary and secondary minima.

Let us neglect the last term $\mathcal{P}_{2m}$ on the right-hand side of Equation (2.3)—we shall return to it in subsequent sections of this chapter—and confine our attention to the first two. In Sec. VI-1 of this chapter we already outlined the way in which we can determine (or estimate) the c_j's from the light changes exhibited between eclipses. To be sure, in doing so we had to split up the light curve into sections which are "certainly unaffected by eclipses", and "possibly affected by them"; and to deal with each consecutively rather than in parallel. However, to segregate the proximity and eclipse effects sequentially would allow the errors entailed in the former step of our analysis to affect the latter, and possibly magnify them in this process.

In the face of such a situation, the question naturally arises as to whether the two could be treated in parallel rather than sequentially—i.e., whether one could "filter out" the photometric proximity effects from the combined light of the system by a *modulation* of the incoming message in the frequency-domain, and suppress (or, at least, lessen) the effects of observational errors on subsequent analysis; and this can indeed be done in the following manner.

In order to do so, let us return to Equation (2.1) which we shall temporarily deprive of its last term on the right-hand side. The two remaining ones (due to proximity and eclipse effects) can, however, be numerically comparable (it is, in general, impossible to say by mere inspection of the light curve which one may turn out to be the larger of the two); and our first task should be to separate them. To this end let us return to Equation (2.1) and multiply both sides of it by the Jacobi polynomial

$$G_n(\lambda, \lambda; x) \equiv {}_2F_1(-n, n + \lambda; \lambda; x) , \quad x \equiv \cos \psi \qquad (2.5)$$

orthogonalized to the interval

$$0 \leq x \leq 1 \qquad (2.6)$$

(corresponding to the range $0 \leq \psi \leq 1$ in phase) with respect to the weight function

$$w(x) = x^{\lambda-1} , \qquad (2.7)$$

in which λ and n are non-negative integers. The polynomials of this type are particular cases of the family studied first by Christoffel (1858); and Radau (1880) or August (1881) investigated their roots for the weight function (2.7) in connection with the Gaussian quadratures (cf. also Section VII-J of Kopal, 1961). As is well known, such polynomials of n-th degree admit of n real and distinct roots in the open interval $0 < x < 1$; and, accordingly, the $G_n(\lambda, \lambda; x)$'s will change sign n times within the same interval—a property of importance for this particular family of Gaussian quadratures. In what follows we propose, however, to employ them for another purpose.

For, having multiplied Equation (2.1) by such polynomials, let us next integrate it with respect to the differential element $d(\cos^\lambda \psi) \equiv dx^\lambda$ between the limits $(0, 1)$ of orthogonality of (2.7). Abbreviating

$$\int_0^1 x^j \, {}_2F_1(-n, n + \lambda; \lambda; x) dx^\lambda \equiv \kappa_j^{(\lambda,n)} , \qquad (2.8)$$

we find that if $j < n$,

$$\kappa_j^{(\lambda,n)} = 0 \qquad (2.9)$$

for any value of λ; while if $j \geq n$,

$$\kappa_j^{(\lambda,n)} = \frac{\lambda}{\lambda + j} \, {}_3F_2(-n, n + \lambda, \lambda + j; \lambda, \lambda + j + 1; 1) . \qquad (2.10)$$

The polynomial ${}_3F_2(1)$ on the right-hand side being Saalschützian (cf., e.g., Bailey, 1935), Equation (2.10) can be rewritten as

$$\kappa_j^{(\lambda,n)} = \frac{\lambda}{\lambda + j} \frac{(-j)_n(-n)_n}{\lambda + j(\lambda)_n(-\lambda - n - j)_n} , \qquad (2.11)$$

where $(x)_n$ stands for the usual Pochhammer product $x(x+1)(x+2) \dots (x+n-1)$. If $j < n$, Equation (2.11) reduces indeed to (2.9); while for $j \geq n$, the $\kappa_j^{(\lambda,n)}$'s become rational fractions tabulated for $\lambda = 1(1)6$, $j = 0(1)5$ and $n = 0(1)5$ in the Appendix.

The polynomial (2.5) represents the "filter" which offers the possibility of annihilating the photometric effects of proximity in the combined light of the system (or suppressing them below the level at which they no longer remain significant). For integrate the simplified equation (2.1) multiplied by it with respect to the differential element $d(\cos^\lambda \psi) \equiv dx^\lambda$ between the limits $(0, 1)$ of its orthogonality: in doing so we find that

$$\int_0^1 \Delta \mathcal{L} \, G_n(\lambda, \lambda; x) dx^\lambda = \sum_{j=1}^n \kappa_j^{(\lambda,n)} c_j + L_1 \int_0^1 \alpha \, G_n(\lambda, \lambda; x) dx^\lambda, \qquad (2.12)$$

where α continues to be given by Eq. (1.4) of Chapter V. If $j < n$, we have seen above that the coefficients $\kappa_j^{(\lambda,n)}$ vanish identically—a fact which reduces Equation (2.1) to

$$\int_0^1 \Delta \mathcal{L} \, G_n(\lambda, \lambda; x) dx^\lambda =$$

$$= L_1 \int_0^1 \alpha \, G_n(\lambda, \lambda; x) dx^\lambda \equiv B_n^{(\lambda)}; \qquad (2.13)$$

the left-hand side of which can be ascertained from the observations, and the right-hand one is a function of the geometrical elements.

In order for this to be exactly so, it is necessary that the degree n of the respective modulating polynomial G_n be greater than the order j of the spherical-harmonic distortion whose photometric proximity effects we wish to suppress. But our modulation process does more; for an inspection of the data given in the Appendix discloses that even if $j \geq n$, the coefficients $\kappa_j^{(\lambda,n)}$ diminish rapidly with increasing values of λ. This is a very fortunate circumstance; for it discloses that also for $n > j$, the c_j's which "slip through" our filter will be factored by very small coefficients; and because of this fact their effects on subsequent analysis will become practically ignorable if not exactly zero.

Next, let us turn our attention to the "modulated moments" $B_n^{(\lambda)}$ of the light curve as defined by Equation (2.13). In order to evaluate it, let us represent the changes $\Delta \mathcal{L}$ of the system by a Fourier series of the form

$$\Delta \mathcal{L} = \frac{1}{2} \sum_{m=0}^M \epsilon_m a_m \cos 2m\psi \qquad (2.14)$$

within the interval $0 < \psi < 90°$—identical with Eq. (4.1) of Chapter V for $\psi_1 = \pi/2$ and $\epsilon_0 = 1$ while for $m > 0$, $\epsilon_m = 2$. If, moreover, we rewrite the Jacobi polynomials (2.5) in ascending powers of x as

$$G_n(\lambda, \lambda; x) = \sum_{i=0}^n \frac{(-n)_i (n + \lambda)_i}{(\lambda)_i \, i!} x^i \qquad (2.15)$$

and insert for $\Delta \mathcal{L}$ from (2.1), Equation (2.13) assumes the form

$$B_n^{(\lambda)} = \frac{1}{2} \sum_{i=0}^n \frac{-n)_i (n + \lambda)_i}{(\lambda + 1)_i \, i!} \sum_{m=0}^M \epsilon_m a_m \int_0^1 \cos 2m\psi \, d(\cos^{i+\lambda} \psi), \qquad (2.16)$$

which can be rewritten, more concisely, as

$$B_n^{(\lambda)} = \sum_{m=0}^{M} \phi_m^{(\lambda,n)} a_m , \tag{2.17}$$

where we have abbreviated

$$\phi_m^{(\lambda,n)} = \frac{\epsilon_m}{2} \sum_{i=0}^{n} \frac{(-n)_i(n+\lambda)_i}{(\lambda+1)_i \, i!} \int_0^1 \cos 2m\psi \, d(\cos^{i+\lambda}\psi). \tag{2.18}$$

Since, moreover,

$$\cos 2m\psi = {}_2F_1(-m,m;;\sin^2\psi) \equiv {}_2F_1(-m,m;;1-x^2), \tag{2.19}$$

it follows that, if $x^2 \equiv y$,

$$\int_0^1 \cos 2m\psi \, d(\cos^{i+\lambda}\psi) = \frac{i+\lambda}{2} \int_0^1 (1-y)^{(1/2)(i+\lambda-2)} {}_2F_1(-m,m;\frac{1}{2};y)dy =$$

$$= {}_3F_2\{-m,m,1;\frac{1}{2},\frac{1}{2}(i+\lambda+2);1\}; \tag{2.20}$$

so that, by (2.17),

$$\phi_m^{(\lambda,n)} = \frac{\epsilon_m}{2} \sum_{i=0}^{n} \frac{(-n)_i(n+\lambda)_i}{(\lambda+1)_i i!} {}_3F_2\{-m,m,1;\frac{1}{2},\frac{1}{2}(i+\lambda+2);1\}. \tag{2.21}$$

The generalized hypergeometric series ${}_3F_2(1)$ on the right-hand side of the preceding equation becomes Saalschützian only if $i+\lambda = 1$—when

$$_3F_2(-m,m,1;\frac{1}{2},\frac{3}{2};1) = \frac{(-\frac{1}{2})_m}{(\frac{3}{2})_m} . \tag{2.22}$$

For all other combinations of λ and i, the weight coefficients $\phi_m^{(\lambda,n)}$ as defined by (2.21) are rational fractions which must be obtained by addition of the individual terms of the summation on the right-hand side of (2.21) as it stands; and the results for $\lambda = 1(1)6$, $m = 0(1)5$ and $n = 0(1)5$ can be found in the Appendix. The reader may note that—regardless of λ—for increasing values of n, more and more ϕ's become zero; and those which do not vanish become increasingly smaller. With the aid of the data collected in that Appendix, the empirical Fourier coefficients a_m characteristic of the respective light curve can, by an appeal to Equation (2.17), be translated into the values of the modulated moments $B_n^{(\lambda)}$ for any desired combination on λ and n.

Are all such combinations independent of each other? This is, in general, *not* the case; for known recursion formulae for the Jacobi polynomials $G_n(\lambda, \lambda; x)$ disclose that, for any value of m,

$$(\lambda + 2n + 1)\phi_m^{(\lambda+1,n)} = (\lambda+1)\{\phi_m^{(\lambda,n)} - \phi_m^{(\lambda,n+1)}\}, \tag{2.23}$$

The reader may also note that (since the modulation polynomials $G_n(\lambda, \lambda; x)$ change sign n times between $x = 0$ and 1), the weight coefficients $\phi_n^{(\lambda, n)}$—and, therefore, the values of the modulated moments $B_n^{(\lambda)}$—can be positive or negative.

What happens if these moments come out to be zero (or insignificant)? Should this be the case, it would follow from Equation (2.13) that the *respective system does not eclipse*; and that the observed light changes are due to the ellipticity (and reflection) effects alone. Should, on the other hand, at least some of the $B_n^{(\lambda)}$'s differ significantly from zero, this would imply that eclipse effects do superpose upon those due to proximity alone; and must be isolated from them.

In order to do so, let us return to Equation (2.13) and focus our attention on the last term on its right-hand side. On insertion in it for $G_n(\lambda, \lambda; x)$ from (2.15) we see that

$$\int_0^1 \alpha G_n(\lambda, \lambda; x) dx^\lambda = \sum_{i=0}^n \frac{(-n)_i(n+\lambda)_i}{(\lambda)_i i!} \int_0^1 \alpha x^i \, dx^\lambda =$$

$$= \sum_{i=0}^n \frac{(-n)_i(n+\lambda)_i}{(\lambda+1)_i i!} \int_0^1 \alpha \, d(\cos^{\lambda+i}\psi) \qquad (2.24)$$

which by the resort to a binomial expansion of $\cos^{i+\lambda}\psi$ in terms of $\sin^2\psi$, and by a suitable choice of the parameters λ and n to annihilate the coefficients $\kappa_j^{(\lambda, n)}$ in (2.12) permits us to establish linear relations between the "modulated" moments $B_n^{(\lambda)}$ and "straight" moments A_{2j} of the light curves, of the form

$$B_n^{(\lambda)} = \sum_{j=1}^\infty \rho_j^{(\lambda, n)} A_{2j} \qquad (2.25)$$

in which advantage was taken of the fact that

$$A_{2j} = L_1 \int_0^{\psi_1} \alpha \, d \sin^{2j}\psi = L_1 \int_0^{\pi/2} \alpha \, d \sin^{2j}\psi \qquad (2.26)$$

(since, for $\psi > \psi_1$, $\alpha = 0$ and, therefore, the upper limit of integration can be restricted from $90°$ to ψ_1); and where

$$\rho_j^{(\lambda, n)} = (-1)^{j+1} \sum_{i=0}^n \frac{(-n)_i(n+\lambda)_i}{(\lambda+1)_i i!} \binom{i+\lambda/2}{j} \qquad (2.27)$$

are appropriate weight coefficients, the numerical values of which are tabulated in the Appendix.

The foregoing equation (2.25) constitutes the fundamental set of linear relations connecting the "modulated" moments $B_n^{(\lambda)}$ of the light curves—the values of which can be obtained, with the aid of Equation (2.15), from the empirically-deduced Fourier coefficients a_m—with the "straight" moments A_{2j} of the light changes which are due to eclipses alone. Provided that the determinant of the coefficients of the A_{2j}'s on the right-hand side of (2.25) does not vanish—i.e., that

$$\| \rho_j^{(\lambda,n)} \| \neq 0 , \tag{2.28}$$

the system (2.25) particularized for a selected combination of the constants λ and n will furnish a unique set of A_{2j}'s, evaluated in terms of the corresponding moments $B_n^{(\lambda)}$ deduced from the observed light changes; and the resulting set of the A_{2j}'s can then be used as a basis for a determination of the elements r_1, r_2, j, and L_1 by the methods of Chapter V.

A suitable choice of disposable constants λ and n to suppress the sum of the terms $\kappa_j^{(\lambda,n)} c_j$ below the "noise level" of our light curve; but not any arbitrary combination will make the determinant of the system different from zero. The condition (2.28) would, for instance, *not* be satisfied for any combination of λ and n for which the absolute terms $B_n^{(\lambda)}$ on the left-hand side of Equation (2.25) would satisfy the equation

$$(\lambda + 2n + 1)\tilde{B}_n^{(\lambda+1)} = (\lambda_1)\{\tilde{B}_n^{(\lambda)} - \tilde{B}_{n+1}^{(\lambda)}\}, \tag{2.29}$$

obtaining from the recursion formula (2.23); and the latter going back to the recursion properties of the modulating polynomials $G_n(\lambda, \lambda; x)$.

In practical cases (see next section), care must be exercised to ensure that this is not the case. Inasmuch as the recursion formula (2.29) for the $B_n^{(\lambda)}$'s entails three terms for λ, $\lambda + 1$, and n, $n + 1$, it cannot be fulfilled if we deploy on the left-hand side terms characterized by λ and $\lambda + 2$ *or* n and $n + 2$. These conditions are sufficient to break the vicious circle represented by (2.29), but not necessarily sufficient to avoid the pitfall of more complicated recursion formulae for the $B_n^{(\lambda)}$'s which may entail more than three terms. Should the investigator be unlucky in his choice to come across one, he will be told so by the vanishing of the determinant of the equations of the form (2.25), which would preclude a unique solution for the A_{2j}'s.

Another point should be made, in this connection, concerning the number of combinations which one can usefully form of the eligible constants λ and n. With increasing values of λ and n (needed to suppress in (2.25) the weighted sum of the c_j's) the total number of such combinations may become very large indeed; but not all of them would be useful. For it is obvious from Equation (2.17) that *the total number of modulated moments* $B_n^{(\lambda)}$ *which are linearly independent of each other cannot exceed the number* $M + 1$ *of terms retained in the Fourier cosine expansion on the right-hand side of* Equation (2.14)—i.e., the number of significantly known coefficients a_m used in the summation on the r.h.s. of (2.17). Any excess of the empirical moments $B_n^{(\lambda)}$ over $M + 1$ would render some of them expressible in terms of others, which would again make the determinant on the left-hand side of (2.28) vanish, and a solution of (2.25) for the A_{2j}'s indeterminate.

Needless to say, the number of the $B_n^{(\lambda)}$'s formed from the observed values of the Fourier coefficients a_m should not only be smaller than $M + 1$, but should also be larger than the number of the moments A_{2j} retained on the right-hand side of (2.25). In order to specify from them the four elements r_1, r_2, i, and L_1 of the system by the methods expounded in Chapter V, a knowledge of at least *four* such moments is prerequisite. We know, moreover, that their numerical values diminish rapidly with increasing values of j—i.e., that

$$A_{2j+2} \ll A_{2j} ; \tag{2.30}$$

and that, beyond a certain limit of the sequence (generally for $j > 4$) higher-order moments become numerically insignificant. Should we, accordingly, break off the summation on the right-hand side of Equation (2.25) after three terms, then for $M > 2$, the system of Equations (2.25) for A_2, A_4, and A_6 becomes overdeterminate, and can be solved for the most probable values of the unknowns by the method of least squares.

VI.3 Photometric Perturbations Arising from Distortion of the Eclipsing Star

In arriving at this stage of our analysis, we have not yet reached the end of the road; for in order to obtain a complete solution of our problem, we still have to incorporate in the process the term $\mathcal{P}_{2m}$, representing photometric perturbations on the r.h.s. of Equation (2.3), which arise from the distortions of the stars during eclipses. The integrands $f_{1,2}^{(n)}$ of the respective expressions are, to be sure, already known to us from Chapter IV; so that, in what follows, it remains for us to evaluate their integrals with respect to the element $d(\sin^{2m} \psi)$ between the limits, 0, 1.

In order to do so, let us first confine our attention to the effects caused by a distorted secondary component eclipsing a spherical mate—a situation realized (or very closely approximated) by eclipsing binaries of semi-detached type, of which Algol can be regarded as a typical example; while the photometric effects caused by eclipses of distorted primary components will be considered in the concluding part of this section.

The photometric effects $f_2^{(n)}$ caused by a distortion of the shadow cylinder cast by the secondary component in the direction of the line of sight are given by Equation (2.21) of Chapter IV, in which (to the first order in surficial distortion) the terms arising from the tides are of the form $l_2^m I_{-1,n}^m$ ($l_2 \equiv \delta$), while the rotational term (factored by $v_2^{(2)}$) can be rewritten as

$$n_1^2 I_{1,n}^0 + n_2^2 I_{-1,n}^2 - \frac{1}{3} I_{-1,n}^0 = \left(\frac{1}{3} - n_2^2\right) I_{-1,n}^0 - (n_1^2 - n_2^2)\frac{r_2}{2\delta\nu} I_{-1,n+2}^1 \tag{3.1}$$

by Eq. (2.28) of Chapter IV, and in which advantage has been taken of the recursion relation

$$(\gamma + 2)(\delta/r_2)\{I_{\beta,\gamma}^m - I_{\beta,\gamma}^{m+2} = (\beta + m + 2)I_{\beta,\gamma+2}^{m+1} - mI_{\beta,\gamma+2}^{m-1}, \tag{3.2}$$

obtained by a combination of Eqs. (2.24)–(2.25) and (2.28), Chapter II; which for $\beta = -1$, $\gamma = n$ and $m = 0, 1, 2$ yields

$$I^2_{-1,n} = I^0_{-1,n} - \left(\frac{r_2}{2\delta}\right)\frac{I^1_{-1,n+2}}{\nu} , \tag{3.3}$$

$$I^3_{-1,n} = I^1_{-1,n} - \left(\frac{r_2}{2\delta}\right)\frac{I^1_{-1,n+2}}{\nu} +$$

$$+ 2\left(\frac{r_2}{2\delta}\right)^2 \frac{I^1_{-1,n+4}}{\nu(\nu+1)} , \tag{3.4}$$

$$I^4_{-1,n} = I^0_{-1,n} - 2\left(\frac{r_2}{2\delta}\right)\frac{I^1_{-1,n+2}}{\nu} +$$

$$+ 3\left(\frac{r_2}{2\delta}\right)^2 \frac{I^0_{-1,n+4}}{\nu(\nu+1)} - 6\left(\frac{r_2}{2\delta}\right)^3 \frac{I^1_{-1,n+6}}{\nu(\nu+1)(\nu+2)} , \tag{3.5}$$

etc., so that

$$3l^2_2 I^2_{-n,1} - I^0_{-1,n} = 2P_2(l_2)I^0_{-1,n} - \frac{3}{2}r_1\left(\frac{r_1}{r_2}\right)^{n+2}\delta\alpha^1_n, \tag{3.6}$$

$$5l^3_2 I^3_{-1,n} - 3l_2 I^1_{-1,n} = 2P_3(l_2)I^1_{-1,n} - \frac{5}{2}\delta r_1\left(\frac{r_1}{r_2}\right)^{n+2}\delta\alpha^1_n +$$

$$+ \frac{5r_1 r_2}{2\nu}\left(\frac{r_1}{r_2}\right)^{n+4}\delta\alpha^1_{n+2} \tag{3.7}$$

and

$$35l^4_2 I^4_{-1,n} - 30l^2_2 I^2_{-1,n} + 3I^0_{-1,n} =$$

$$= 8P_4(l_2)I^0_{-1,n} - 5r_2\left(\frac{r_1}{r_2}\right)^{n+2}(7\delta^2 - 3)\delta\alpha^1_n +$$

$$+ \frac{105}{4\nu}\delta r_1 r_2\left(\frac{r_1}{r_2}\right)^{n+4}\delta\alpha^1_{n+2} -$$

$$- \frac{105 r_1 r^2_2}{4\nu(\nu+1)}\left(\frac{r_1}{r_2}\right)^{n+6}\delta\alpha^1_{n+4} . \tag{3.8}$$

If, moreover, we recall that, by (2.5) of Chapter IV,

$$w^{(4)}_2 \cong r_2 w^{(3)}_2 \cong r^2_2 w^{(2)}_2 \tag{3.9}$$

(an approximation sufficient for centrally-condensed stars) while, by (1.14)–(1.15) of that chapter

$$n^2_2 = n^2_0(l^{-2}_2 - 1) \tag{3.10}$$

and

$$n_1^2 - n_2^2 = 1 - 2(n_0/l_2)^2 + n_0^2 \; ; \tag{3.11}$$

where, for eclipsing binaries, the direction cosine n_0^2 can also be regarded as a small quantity (very small for totally-eclipsing systems); though the same need not be true of the ratios n_0/l_2.

As a result, to the accuracy equivalent to first-order in surficial distortion, Eq. (2.21) of Chapter IV reduces to

$$\left(\frac{r_1}{r_2}\right)^{n+2} f_2^{(n)} = -\left\{\left[\frac{1}{3} + n_0^2 - \left(\frac{n_0}{\delta}\right)^2\right] I_{-1,n}^0 - \right.$$

$$- \left[1 - 2\left(\frac{n_0}{\delta}\right)^2 + n_0^2\right]\left(\frac{r_2}{2\delta\nu}\right) I_{-1,n+2}^1 \left.\right\} v_2^{(2)} +$$

$$+ (3\delta^2 - 1)I_{-1,n}^0 \, w_2^{(2)} +$$

$$+ \frac{3}{4}\left\{I_{-1,n}^0 - 4\left(\frac{\delta}{r_2}\right)I_{-1,n}^1 - 2\left(\frac{\delta}{\nu r_2}\right)I_{-1,n+2}^1\right\} w_2^{(4)} +$$

$$+ \cdots , \tag{3.12}$$

where $2\nu \equiv n + 2$.

In order to ascertain the photometric perturbations $\mathcal{P}_{2m}$ on the right-hand side of Eq. (2.3), all we need to do is to evaluate the integrals of $\delta^{2J} I_{-1,n}^0$ and $\delta^{2(J-1)}(\delta I_{-1,n}^1)$, $J = 0$ and ± 1, with respect to $d(\sin^{2m}\psi) \equiv m(\delta^2 - \delta_0^2)^{m-1}$ $(1 - \delta_0^2)^{-m} d\delta^2$ between $\delta_0 \equiv n_0$ and δ_1 and form their sum as indicated by the foregoing Equation (3.12). For any type of eclipse, the functions $I_{-1,n}^0$ and $\delta I_{-1,n}^1$ are known to us already from Equation (3.9) of Chapter III; but in order to prepare them for subsequent integration, we find it of advantage to change over the arguments of the Jacobi polynomials involved in them from a to $b = 1 - a$ and from c^2 to $1 - c^2$. This can be done by means of the identities

$$G_{j+1}(\nu + \kappa, \nu + \kappa + 1; a) \equiv G_j(\nu + \kappa + 2, \nu + \kappa + 1; a), \tag{3.13}$$

$\kappa = 0, 1$ and

$$(-1)^j (j + 1)! \, G_j(\nu + 2, 2; a) \equiv (\nu + 1)_j \, G_j(\nu + 2, \nu + 1; b); \tag{3.14}$$

while

$$(-1)^j j! \, G_j(\nu + 2, 1; c^2) \equiv (\nu + 2)_j \, G_j(\nu + 2, \nu + 2; 1 - c^2); \tag{3.15}$$

by virtue of which Equation (3.9) of Chapter III can be rewritten as

$$I_{-1,n}^\kappa = \left(\frac{a}{b}\right)^{2\nu}\left(\frac{c}{b}\right)^\kappa \frac{(1 - c^2)^\nu}{\nu} \sum_{j=0}^\infty (j + 1)_\kappa (\nu + j + 1)_\kappa \times$$

$$\times (\nu + 2j + \kappa + 1)\{bG_j(\nu + \kappa + 1, \kappa + 1; b)\}^2 \times$$

$$\times G_j(\nu + \kappa + 1, \nu + 1; 1 - c^2), \tag{3.16}$$

and, accordingly,

$$\int_0^{\psi_1} \delta^{2(J-\kappa)} (\delta^\kappa I^\kappa_{-1,n}) d(\sin^{2m}\psi) =$$

$$= \frac{m}{\nu r_2} \left(\frac{a}{r_2}\right)^{2\nu} \int_{j=0}^\infty (j+1)_\kappa (\nu+j+1)_\kappa (\nu+2j+\kappa+1) \times$$

$$\times \{G_j(\nu+\kappa+1, \kappa+1; b)\}^2 \int_{\delta_0^2}^{\delta_1^2} \delta^{2J}(\delta^2 - \delta_0^2)^{m-1}(1-c^2)^\nu \times$$

$$\times G_j(\nu+\kappa+1, \nu+1; 1-c^2)d\delta^2, \tag{3.17}$$

where the parameters a, b, c continue to be defined by Eqs.(2.3)–(2.5) of Chapter III.

In order to normalize the limits of integration on the r.h.s. of the preceding equation, let us introduce the variable u as defined by Equation (2.2) of Chapter V, in accordance with which

$$\delta^2 = \sin^2\psi \sin^2 j + \cos^2 j \equiv n_0^2 + m_{01}^2 u, \tag{3.18}$$

m_{01}, n_0 being the direction cosines of the line of sight, as defined by Eqs. (1.23) of Chapter IV at the moment of first contact of the eclipse (when $\psi = \psi_1$); so that

$$\delta_1^2 - \delta^2 = m_{01}^2(1-u) \text{ and } \delta^2 - \delta_0^2 = m_{01}^2 u; \tag{3.19}$$

Since, moreover, by definition

$$G_j(\kappa+\nu+1, \kappa+1; 1-c^2) = \sum_{i=0}^j (-1)^i \binom{j}{i} \frac{(j+\kappa+\nu+1)_i}{(\kappa+1)_i}(1-c^2), \tag{3.20}$$

where

$$1 - c^2 = (1-c_0^2)(1-u) \equiv n_{11}^2(1-u) \tag{3.21}$$

and $n_{11}^2 \equiv (1-c_0^2)$ signifies the direction cosine n_1 (as given by Eq.(1.14), Chapter IV) at the moment of first contact (when $\psi = \psi_1$).

If so, then for $J = 0$,

$$\int_{\delta_0^2}^{\delta_1^2} (\delta^2 - \delta_0^2)^{m-1}(1-c^2)^\nu G_j(\kappa+\nu+1, \kappa+1; 1-c^2)d\delta^2 =$$

$$= (m_{01})^{2m}(n_{11})^{2\nu} \sum_{i=0}^j (-1)^i \binom{j}{i} \frac{(j+\kappa+\nu+1)_i}{(\kappa+1)_i} \int_0^1 u^{m-1}(1-u)^{\nu+i}du =$$

$$= (m_{01})^{2m}(n_{11})^{2\nu} \sum_{i=0}^j (-1)^i \binom{j}{i} \frac{(j+\kappa+\nu+1)_i}{(\kappa+1)_i} B(m, i+\nu+1), \tag{3.22}$$

where $B(m, i+\nu+1)$ stands for the (complete) beta-function of the respective arguments.

On the basis of all the preceding results we eventually establish that

$$\int_0^{\psi_1} \delta^\kappa I_{-1,n}^\kappa d(\sin^{2m}\psi) = r_2^\kappa \sin^{2m}\psi_1 \left(\frac{r_1}{r_2}\right)^{2\nu} \times$$

$$\times (1 - c_0^2)^\nu B(m+1, \nu) \sum_{j=0}^{\infty}(j+1)_\kappa(\nu+j+1)_\kappa \times$$

$$\times G_{j+\kappa}(\nu+\kappa, m+\nu+1; 1-c_0^2). \tag{3.23}$$

For $J = 1$,

$$\int_{\delta_0^2}^{\delta_1^2}(\delta^2 - \delta_0^2)^{m-1}(1-c^2)^\nu G_j(\kappa+\nu+1, \kappa+1; c^2)\delta^2 d\delta^2 =$$

$$= (m_{01})^{2m} n_{11}^{2\nu} \sum_{i=0}^{j} \frac{(-1)^i j!(j+\kappa+\nu+1)_i}{(j-i)!(\kappa+1)_i} n_{11}^{2i} \times$$

$$\times \{m_{01}^2 B(m+1, i+\nu+1) + n_0^2 B(m, i+\nu+1)\}; \tag{3.24}$$

while for $J = -1$ (met only in terms representing rotational distortion, factored by n_0^2),

$$n_0^2 \int_{\delta_0^2}^{\delta_1^2}(\delta^2 - \delta_0^2)^{m-1}(1-c^2)^{2\nu} G_j(\kappa+\nu+1, \kappa+1; 1-c^2)\delta^2 d\delta^2 =$$

$$= (m_{01})^{2m} n_{11}^{2\nu} \sum_{i=0}^{j} \frac{(-1)^i j!(j+\kappa+\nu+1)_i}{(\kappa+1)_i} n_{11}^{2i} \times$$

$$\times \int_0^1 \frac{u^{m-1}(1-u)^{\nu+i} du}{1+(m_{01}/n_0)^2 u}; \tag{3.25}$$

where the remaining integral can be recognized as a representation of the hypergeometric series of the form

$$\int_0^1 \frac{u^{m-1}(1-u)^{\nu+i} du}{1+(m_{01}/n_0)^2 u} =$$

$$= B(m, i+\nu+1)m_{11}^2\, _2F_1(1, \nu+i+1; m+\nu+1; n_{11}^2) =$$

$$= B(m, i+\nu+1)m_{11}^2 \sum_{j=1}^{\infty} \frac{(\nu+i+1)_j}{(m+\nu+i+1)_j} n_{11}^{2j}, \tag{3.26}$$

where m_{11} and n_{11} stand for the particular values of the direction cosines m_1 and n_1 (as given by Eqs. (1.14), Chapter IV) at the phase angle ψ_1.

By this all expressions necessary to formulate the photometric perturbations $\mathcal{P}_{2m}$ as defined by Eq. (2.3) of this chapter have been evaluated, valid for the eclipses of *any* type—total, annular or partial (occultations or transits). Should, however, the eclipses become *total*, the foregoing anaysis admits of far-reaching simplification; for, in such a case, the products $\delta^{2\mu}(\delta I^{\kappa}_{-1,n})$ are to be integrated with respect to $d(\sin^{2m}\psi)$ between $\delta_{1,2} = r_2 \pm r_1$ and (cf. Appendix A of Kopal, 1987)

$$\int_{\delta_2^2}^{\delta_1^2} \delta^{2\mu}(\delta I^{\kappa}_{-1,n})d\delta^2 = r_2^{2(\mu+1)+\kappa}\frac{k^{2\nu}}{\nu}\times$$
$$\times \, {}_2F_1(-\mu, -\kappa-\mu; \nu+1; k^2)\,, \tag{3.27}$$

$k \equiv r_1/r_2$; which for $\kappa = 0$, 1 and zero or integral values of μ reduces to the Jacobi polynomials in k^2.

Combining Equation (2.4) with Eqs. (3.12) and (3.17) we eventually establish that

$$p^2\mathcal{P}_2 = \frac{1}{3}v_2^{(2)} - w_2^{(2)} + \frac{3}{4}w_2^{(4)} -$$
$$- \frac{3}{4}X_1\left(\frac{1}{3}v_2^{(2)} - w_2^{(4)}\right) - q^2v_2^{(2)} +$$
$$+ (1-\frac{1}{4}X_1)q^2r_2^2v_2^{(2)}\,, \tag{3.28}$$

$$p^4\mathcal{P}_4 = \frac{1}{3}\{2(1-q^2)(1-3q^2) - \frac{1}{2}(1-9q^2)X_1\}v_2^{(2)} +$$
$$+ q^2\{2(1-q^2) + \frac{1}{2}(1+q^2)X_1\}r_2^2v_2^{(2)} -$$
$$- \{2(1-q^2) + X_1\}w_2^{(2)} +$$
$$+ \frac{3}{4}\{2(1-q^2) + (7-2q^2)X_1 + \frac{4}{3}X_2\}w_2^{(4)}, \tag{3.29}$$

and

$$p^6\mathcal{P}_6 = \frac{2}{3}\Big\{(1-q^2)^2(1-3q^2)-$$
$$- \left[\frac{15}{4}(1-q^2)^2 - 7(1-q^2) + 2\right]X_1 - \frac{1}{6}X_2\Big\}v_2^{(2)} +$$
$$+ 2q^2\Big\{(1-q^2)^2 - \frac{1}{4}[(1-q^2)^2 - 4(1-q^2) - 4]X_1 - \frac{1}{4}X_2\Big\}r_2^2v_2^{(2)} -$$
$$+ \frac{3}{2}\Big\{(1-q^2)^2 + [(1-q^2)^2 + 5(1-q^2) + 1]+$$
$$+ \left[\frac{4}{3}(1-q^2) + 5\right]X_2 + \frac{1}{2}X_3\Big\}w_2^{(4)}, \tag{3.30}$$

where we have abbreviated

$$p \equiv \frac{\sin j}{r_2}, \quad q \equiv \frac{\cos j}{r_2} \tag{3.31}$$

and

$$X_j \equiv (j+1)! \, k^{2j} \sum_{n=0}^{N} \frac{C^{(n)}}{(\nu)_{j+1}}. \tag{3.32}$$

The strategy for incorporating the photometric effect of the distortion of the shadow cylinder of the eclipsing component cast on its mate can now be outlined as follows.

1. First, let us expand the observed variation of light of our eclipsing system between conjunction and quadrature (i.e., between the phase angles ψ and $90°$) in a Fourier series of the form (2.14), and determine its coefficients a_j by the method of least squares; the areas $\overline{A}_{2m}$ subtended between the curves $\mathcal{L}(\pi/2) - \mathcal{L}(\psi)$ and $\sin^{2m} \psi$ and $\psi = 0$ can be estimated by planimetry (see Fig. VI.1); or analytically, in terms of the A_j's, by Equations (4.3)–(4.6) of Chapter V in which we have set $\psi = \pi/2$: i.e.,

$$\overline{A}_2 = \frac{1}{2}a_0 + a_1 + a_2 + a_3 + a_4 + \cdots , \tag{3.33}$$

$$\overline{A}_2 = \frac{1}{2}a_0 \qquad -\frac{1}{3}a_2 \qquad -\frac{1}{15}a_4 + \cdots , \tag{3.34}$$

$$\overline{A}_4 = \frac{1}{2}a_0 - \frac{1}{3}a_1 - \frac{1}{3}a_2 + \frac{1}{5}a_3 - \frac{1}{15}a_4 + \cdots , \tag{3.35}$$

$$\overline{A}_6 = \frac{1}{2}a_0 - \frac{1}{2}a_1 - \frac{1}{5}a_2 + \frac{3}{10}a_3 - \frac{1}{7}a_4 + \cdots , \tag{3.36}$$

etc.

2. Next, let us turn to Equation (2.3), neglect $\mathcal{P}_{2m}$; and if it is possible to determine the coefficients c_j by the method of Sec. VI-1, transpose the c_j's to the left-hand side to isolate a first approximation to the moments of the light curve A_{2m}.

3. An alternative way to obtain these is to "modulate" the entire observed light curve by the Jacobian polynomials to obtain (by planimetry or otherwise) the "modulated" moments $B_n^{(\lambda)}$ as defined by Eq. (2.13), and then solve for the A_{2j}'s from Equations (2.25).

4. Once a satisfactory set of these moments (freed from the effects of distortion) has been obtained, a solution for the (approximate) geometrical elements r_1, r_2, j and L_1 can be carried out by the methods developed in Chapter V. With their aid, photometric perturbations $\mathcal{P}_{2m}$ can be numerically estimated by the formulae developed in Sec. VI-3, transposed to the left-hand side of Equations (2.3) or (2.24); and the solution for the elements repeated until the adopted and resulting elements are no longer significantly different.

VI.4   Effects Caused by Distortion of the Star Undergoing Eclipse

The process described so far in this section is not yet sufficient to free the observed light changes of close eclipsing systems from all photometric effects of distortion; and the bulk of them remains, in fact, still to be investigated: namely, those arising from the distortion of the eclipsed star. The main reason is the fact that while the effects caused by the distortion of the eclipsing star are purely geometrical (i.e., depend only on the shape of its limb), a distortion of the star undergoing eclipse will affect its light not only through the geometry of its figure, but also by its effect on the distribution of brightness of the star's apparent disc (or a crescent thereof). Therefore, it is only to be expected that the number of terms still to be considered should be larger than that considered in Sec. VI-3; and this expectation will indeed not be disappointed. For the contribution of the primary component to total photometric perturbations $\mathcal{P}_{2m}$, as given by Eq. (2.4), will be of the form

$$\mathcal{P}_{2m} = \sum_{n=0}^{N} C^{(n)} \int_{0}^{\psi_1} f_1^{(n)} d(\sin^{2m} \psi) \equiv \sum_{n=0}^{N} \frac{C^{(n)}}{\nu} \mathcal{P}_{2m}^{(n)} \tag{4.1}$$

where the integrand $f_1^{(n)}$ has already been detailed in Equations (3.31)–(3.43) of Chapter IV. Moreover, if (by analogy of Equation (3.9) in Sec. VI-3) we set, in those equations

$$w_1^{(4)} \cong r_1 w_1^{(3)} \cong r_1^2 w_1^{(2)} \tag{4.2}$$

and

$$l_0 = \sqrt{1 - \delta^2} = 1 - \frac{r_1^2}{2} \left(\frac{\delta}{r_1} \right)^2 + \cdots , \tag{4.3}$$

then the tidal harmonic distortion terms will reduce to

$$f_1^{(n)} = -\frac{1}{2} \{ 3(\beta_2 - n)\alpha_{n+2}^0 - (\beta_2 - 2n)\alpha_n^0 + n\alpha_{n-2}^0 \} w_1^{(2)}$$

$$- 3\{\beta_2 - n\}\delta\alpha_n^1 w_1^{(2)} +$$

$$+ \frac{3}{4} \{ 3(\beta_2 - n)\alpha_{n-2}^0 - (\beta_2 - 2n)\alpha_n^0 + n\alpha_{n-2} \} \delta^2 w_1^{(2)} -$$

$$- \frac{3}{4} \left\{ \frac{\beta_2 - 6\nu - 2}{\nu} \left[\frac{r_1}{\delta} \alpha_{n+2}^1 - \Im_{-1,n+2}^0 \right] \right.$$

$$+ 2 \left[\frac{r_1}{\delta} \alpha_n^1 - \Im_{-1,n}^0 \right] \Big\} \delta^2 w_1^{(2)} +$$

$$+ \frac{1}{4} \left\{ [6\beta_3 - 5(5n + 3)]\alpha_{n+1}^0 - \right.$$

$$- 5[2\beta_3 - 5n - 5]\alpha^0_{n+3}\Big\} w^{(3)}_1 -$$

$$- \frac{3}{4}\Big\{5(\beta_3 - 4n - 10)\alpha^1_{n+2} + (\beta_3 + 11n + 18)\alpha^1_n - n\alpha^1_{n-2}\Big\} \delta w^{(3)}_1 -$$

$$- \frac{1}{8}\Big\{35(\beta_4 - 3n)\alpha^0_{n+4} - 15(2\beta_4 - 9n)\alpha^0_{n+2}$$

$$+ 3(\beta_4 - 9n)\alpha^0_n - 3n\alpha^0_{n-2}\Big\} w^{(4)}_1. \tag{4.4}$$

In order to integrate these expressions between $(0, \psi_1)$ with respect to $d\sin^{2m}\psi$ to obtain the corresponding contributions $\mathcal{P}^{(n)}_{2m}$, the following types of integrands are encountered repeatedly: namely, α^0_n, δ^2_n and $(r_1/\delta)\alpha^1_n - \mathfrak{F}^0_{-1,n}$. The expansions of the functions α^0_n, $\delta\alpha^1_n \equiv (r_2/r_1)^{n+3}(\delta/\nu)I^1_{-1,n+2}$ and $\mathfrak{F}^0_{-1,n}$, valid for any type of eclipse when $n \geq -1$, have, however, already been established by Equations (3.11) and (3.13) of Chapter III; whence it follows that

$$\frac{r_1}{\delta}\alpha^1_n - \mathfrak{F}^0_{-1,n} = \frac{(ac)^2(1 - c^2)^\nu}{\nu}\sum_{j=0}^{\infty}(j + 1) \times$$

$$\times (\nu + j + 2)(\nu + 2j + 3)\{bG_j(\nu + 3, 2; b)\}^2 \times$$

$$\times G_j(\nu + 3, \nu + 1; 1 - c^2) ; \tag{4.5}$$

and their contributions to $\mathcal{P}_{2m}$ have also been established before: for, in accordance with Eq. (2.3) of Chapter V

$$\int_0^{\psi_1} \alpha^0_n \, d(\sin^{2m}\psi) = m!\sin^{2m}\psi_1\frac{(1 - c^2_0)^{\nu+1}}{(\nu)_{m+2}} \times$$

$$\times \sum_{j=0}^{\infty}(j + 1)(\nu + j + 1)(\nu + 2j + 2) \times \tag{4.6}$$

$$\times \{bG_j(\nu + 2, 2; b)\}^2 G_j(\nu + 2, \nu + m + 2; 1 - c^2_0),$$

which for total eclipses reduces to polynomials similar to Eqs. (1.25)–(1.27) of that chapter; while

$$\int_0^{\psi_1} \left(\frac{r_1}{\delta}\alpha^1_n - \mathfrak{F}^0_{-1,n}\right) d(\sin^{2m}\psi) = (a\sin^m\psi_1)^2\times$$

$$\times (1 - c^2_0)^\nu B(m + 1, \nu)\sum_{j=0}^{\infty}(j + 1)(\nu + j + 2)(\nu + 2j + 3) \times$$

$$\times \{bG_j(\nu + 3, 2; b)\}^2 G_{j+1}(\nu + 1, m + \nu + 1; 1 - c^2_0); \tag{4.7}$$

for any type of eclipse; but if the latter are total,

$$p^{2m}\int_0^{\psi_1} \delta^{2\mu}\Big\{\frac{r_1}{\delta}\alpha^1_n - \mathfrak{F}^0_{-1,n}\Big\} d(\sin^{2m}\psi) = -\frac{m!}{\nu(\nu + 1)}\times \tag{4.8}$$

$$\times (r_2^\mu k)^2 \sum_{j=0}^{m-1} (j - m - \mu + 1)\, {}_2F_1(j - m - \mu + 3,\, j - m - \mu + 2;\, \nu + 2;\, k^2)$$

where, as before, $k \equiv r_1/r_2$ and $p \equiv \sin j/r_2$.

With the aid of the results state in the preceding part of this section, the contributions $f_1^{(n)}$ to the photometric perturbations $\mathcal{P}_{2m}$ as defined by Eq. (4.1) can be expressed as

$$\mathcal{P}_2^{(n)} = -\frac{h(\beta_2 + 4)}{6(h + 3)} C_3\, v_1^{(2)} -$$

$$-\frac{2}{3} \frac{\beta_2 - h + 1}{h + 3} r_1\, v_1^{(2)} \cot^2 j +$$

$$+\frac{\beta_2 - 2h}{3(h + 3)(h + 5)} C_1\, v_1^{(2)} -$$

$$-\left\{ \frac{h(\beta_2 + 4)}{h + 3} w_1^{(2)} + \frac{(h - 1)(h + 1)}{(h + 2)(h + 4)}(\beta_3 + 10) w_1^{(3)} + \right.$$

$$\left. +\frac{h(h - 1)}{(h + 3)(h + 5)}(\beta_4 + 18) w_1^{(4)} + \cdots \right\} C_3 -$$

$$-\frac{3}{p^2} \left\{ \frac{2(\beta_2 - h + 1)}{h + 3} w_1^{(3)} + \right.$$

$$\left. +\frac{(3h + 5)\beta_3 - (h - 3)\beta_2 - 5(h + 1)^2 - 4}{(h + 3)(h + 5)} w_1^{(4)} \right\} +$$

$$+\frac{3}{4} \left\{ \frac{h(\beta_2 + 4)}{h + 3}(r_2^4 - n_0^4) \right\} w_1^{(2)} + \cdots , \tag{4.9}$$

$$\mathcal{P}_4^{(n)} = -\frac{2}{3} \left\{ \frac{h(\beta_2 + 4)}{4(h + 3)} C_3^2 + \frac{(h - 4)\beta_2 + 2q^2 + 6(h + 2)}{(h + 4)(h + 5)} C_2^2 \right\} v_1^{(2)} -$$

$$-C_3^2 \left\{ \frac{h(\beta_2 + 4)}{h + 3} w_1^{(2)} + \frac{(h - 1)(h + 1)}{(h + 2)(h + 4)}(\beta_3 + 10) w_1^{(3)} + \cdots \right\}$$

$$-4C_2^2 \left\{ \frac{h - 1)\beta_2 + 12}{(h + 3)(h + 5)} w_1^{(2)} + \right.$$

$$\left. +\frac{(h + 1)[2(h - 4)\beta_3 + 15(3h - 2)]}{2(h + 2)(h + 4)(h + 6)} w_1^{(3)} + \cdots \right\}$$

$$-\frac{12(\beta_2 - h + 1)}{p^4(h + 3)} \left\{ 1 - q^2 + \frac{4k^2}{h + 5} \right\} w_1^{(3)} + \cdots , \tag{4.10}$$

$$\mathcal{P}_6^{(n)} = -\left\{\frac{h(\beta_2+4)}{6(h+3)}C_3^3 + 2\frac{(h-7)\beta_2+4(5h+1)}{(h+3)(h+5)}C_2^2C_3+ \right.$$

$$\left. +4\frac{(h-8)\beta_2+24h}{(h+3)(h+5)(h+7)}C_1C_2^2 + \cdots \right\}v_1^{(2)} -$$

$$-\left\{\frac{h(\beta_2+4)}{h+3}C_3^3 + 12\frac{(h-1)\beta_2+4(2h+1)}{(h+3)(h+5)}C_2^2C_3+ \right.$$

$$\left. +24\frac{(h-2)\beta_2+12(h+1)}{(h+3)(h+5)(h+7)}C_1C_2^2 + \cdots \right\}w_1^{(2)} + \cdots \qquad (4.11)$$

etc., where $h \equiv n+1$ and β_j continues to be given by Eq. (3.45) of Chapter IV; $C_{1,2,3}$ by Eqs. (2.11)–(2.13), Chapter V; and p, q by Eq. (3.31) of this chapter.

An incorporation of photometric perturbations represented by Eq. (4.4) can be accomplished in the same way as we did with those arising from the secondary component in Sec.VI-3; and once we have done so, the solution of our problem can be regarded as complete.

Appendix

Tables of the coefficients $\kappa_j^{(\lambda,n)}$ in Equation (2.12) as defined by Equation (2.11) for $\kappa = 1(1)6$, $j = 0(1)5$ and $n = 0(1)5$.

$$\lambda = 1$$

n $\diagdown$ j	0	1	2	3	4	5
0	1	$\frac{1}{2}$	$\frac{1}{3}$	$\frac{1}{4}$	$\frac{1}{5}$	$\frac{1}{6}$
1	0	$-\frac{1}{6}$	$-\frac{1}{6}$	$-\frac{3}{20}$	$-\frac{2}{15}$	$-\frac{5}{42}$
2	0	0	$\frac{1}{30}$	$\frac{1}{20}$	$\frac{2}{35}$	$\frac{5}{84}$
3	0	0	0	$-\frac{1}{140}$	$-\frac{1}{70}$	$-\frac{5}{252}$
4	0	0	0	0	$\frac{1}{630}$	$\frac{1}{252}$
5	0	0	0	0	0	$-\frac{1}{2772}$

$$\lambda = 2$$

n $\diagdown$ j	0	1	2	3	4	5
0	1	$\frac{2}{3}$	$\frac{1}{2}$	$\frac{2}{5}$	$\frac{1}{3}$	$\frac{2}{7}$
1	0	$-\frac{1}{12}$	$-\frac{1}{10}$	$-\frac{1}{10}$	$-\frac{2}{21}$	$-\frac{5}{56}$
2	0	0	$\frac{1}{90}$	$\frac{2}{105}$	$\frac{1}{42}$	$\frac{5}{189}$
3	0	0	0	$-\frac{1}{560}$	$-\frac{1}{252}$	$-\frac{1}{168}$
4	0	0	0	0	$\frac{1}{30.105}$	$\frac{1}{11.105}$
5	0	0	0	0	0	$-\frac{1}{4.7.9.11}$

$$\lambda = 3$$

n $\diagdown$ j	0	1	2	3	4	5
0	1	$\frac{3}{4}$	$\frac{3}{5}$	$\frac{1}{2}$	$\frac{3}{7}$	$\frac{3}{8}$
1	0	$-\frac{1}{20}$	$-\frac{1}{15}$	$-\frac{1}{14}$	$-\frac{1}{14}$	$-\frac{5}{72}$
2	0	0	$-\frac{1}{210}$	$\frac{1}{112}$	$\frac{1}{84}$	$\frac{1}{72}$
3	0	0	0	$-\frac{1}{1680}$	$-\frac{1}{700}$	$-\frac{1}{400}$
4	0	0	0	0	$\frac{1}{3.5.7.10.11}$	$\frac{1}{5.8.9.11}$
5	0	0	0	0	0	$-\frac{1}{7.8.9.11.13}$

$$\lambda = 4$$

n \ j	0	1	2	3	4	5
0	1	$\frac{4}{5}$	$\frac{2}{3}$	$\frac{4}{7}$	$\frac{1}{2}$	$\frac{4}{9}$
1	0	$-\frac{1}{30}$	$-\frac{1}{21}$	$-\frac{3}{56}$	$-\frac{1}{18}$	$-\frac{1}{18}$
2	0	0	$\frac{1}{420}$	$\frac{1}{210}$	$\frac{1}{150}$	$\frac{4}{495}$
3	0	0	0	$-\frac{1}{4200}$	$-\frac{1}{1650}$	$-\frac{1}{990}$
4	0	0	0	0	$\frac{1}{5.7.9.10.11}$	$\frac{4}{5.7.9.11.13}$
5	0	0	0	0	0	$\frac{1}{6^2 7^2 11.13}$

$$\lambda = 5$$

n \ j	0	1	2	3	4	5
0	1	$\frac{5}{6}$	$\frac{5}{7}$	$\frac{5}{8}$	$\frac{5}{9}$	$\frac{1}{2}$
1	0	$-\frac{1}{42}$	$-\frac{1}{28}$	$-\frac{1}{24}$	$-\frac{2}{45}$	$-\frac{1}{22}$
2	0	0	$\frac{1}{756}$	$\frac{1}{360}$	$\frac{2}{495}$	$\frac{1}{198}$
3	0	0	0	$-\frac{1}{5.6.7.8.11}$	$-\frac{1}{5.7.9.11}$	$-\frac{1}{11.13.14}$
4	0	0	0	0	$\frac{1}{7.9.10.11.13}$	$\frac{1}{4.7^2 11.13}$
5	0	0	0	0	0	$-\frac{1}{6.7.9.11.13.14}$

$$\lambda = 6$$

n \ j	0	1	2	3	4	5
0	1	$\frac{6}{7}$	$\frac{3}{4}$	$\frac{2}{3}$	$\frac{3}{5}$	$\frac{6}{11}$
1	0	$-\frac{1}{56}$	$-\frac{1}{36}$	$-\frac{1}{30}$	$-\frac{2}{55}$	$-\frac{5}{132}$
2	0	0	$\frac{1}{1260}$	$\frac{2}{1155}$	$\frac{1}{385}$	$\frac{10}{3003}$
3	0	0	0	$-\frac{1}{10.11.12.14}$	$-\frac{3}{10.11.13.14}$	$-\frac{15}{8.11.13.49}$
4	0	0	0	0	$\frac{1}{11.13.30.49}$	$\frac{1}{11.13.18.49}$
5	0	0	0	0	0	$-\frac{1}{7.9.11.13.14.16}$

Tables of the weight coefficients $\phi_j^{(\lambda,n)}$ in Equation (2.17), as defined by Equation (2.21) for $\lambda = 1(1)6$, $j = 0(1)5$ and $n = 0(1)5$.

$$\lambda = 1$$

n	m 0	1	2	3	4	5
0	$\frac{1}{2}$	$-\frac{1}{3}$	$-\frac{1}{15}$	$-\frac{1}{35}$	$-\frac{1}{63}$	$-\frac{1}{99}$
1	0	$-\frac{1}{3}$	$\frac{4}{15}$	$-\frac{1}{35}$	$\frac{16}{315}$	$-\frac{1}{99}$
2	0	$\frac{1}{15}$	$\frac{4}{21}$	$-\frac{5}{21}$	$\frac{16.17}{5.7.9.11}$	$-\frac{673}{7.9.11.13}$
3	0	0	$-\frac{4}{35}$	$-\frac{8}{3.5.7}$	$\frac{16.41}{5.7.9.11}$	$-\frac{8.107}{7.9.11.13}$
4	0	0	$\frac{4}{315}$	$\frac{8.19}{3.5.7.11}$	$-\frac{16}{7.11.13}$	$-\frac{8.47}{3.7.11.13}$
5	0	0	0	$-\frac{8}{231}$	$-\frac{17.64}{7.9.11.13}$	$\frac{8}{99}$

$$\lambda = 2$$

n	m 0	1	2	3	4	5
0	$\frac{1}{2}$	0	$-\frac{1}{3}$	0	$-\frac{1}{15}$	0
1	0	$-\frac{1}{5}$	$\frac{4}{105}$	$\frac{11}{105}$	$-\frac{16}{11.105}$	$\frac{97}{11.13.21}$
2	0	$\frac{1}{45}$	$\frac{32}{315}$	$-\frac{17}{315}$	$-\frac{8.16}{5.7.9.11}$	$\frac{61}{7.9.11.13}$
3	0	0	$-\frac{2}{63}$	$\frac{4}{77}$	$\frac{8.17^2}{5.7.9.11.13}$	$\frac{4.17}{7.9.11.13}$
4	0	0	$\frac{4}{5.5.7.9}$	$\frac{64}{5.5.7.11}$	$\frac{16.59}{5.7.9.11.13}$	$-\frac{64.29}{5.7.9.11.13}$
5	0	0	0	$-\frac{8}{9.11.13}$	$-\frac{64}{3.5.11.13}$	$-\frac{8}{3.11.13.17}$

$$\lambda = 3$$

n	m 0	1	2	3	4	5
0	$\frac{1}{2}$	$\frac{1}{5}$	$-\frac{13}{35}$	$-\frac{11}{105}$	$-\frac{61}{11.105}$	$-\frac{97}{11.13.21}$
1	0	$-\frac{2}{15}$	$-\frac{4}{105}$	$\frac{2}{21}$	$\frac{16}{11.105}$	$\frac{46}{11.13.21}$
2	0	$\frac{1}{105}$	$\frac{2}{35}$	$-\frac{1}{11.105}$	$-\frac{8.71}{11.13.105}$	$-\frac{4}{11.13.21}$
3	0	0	$-\frac{2}{5.35}$	$-\frac{4.41}{55.105}$	$\frac{8.19}{105.143}$	$\frac{4.61}{105.143}$
4	0	0	$\frac{4}{55.105}$	$\frac{8.101}{5.105.143}$	$\frac{16}{11.105}$	$-\frac{8.349}{17.105.143}$
5	0	0	0	$-\frac{32}{7.21.143}$	$-\frac{64}{49.143}$	$-\frac{8.12}{7.17.143}$

$$\lambda = 1$$

n \ m	0	1	2	3	4	5
0	$\frac{1}{2}$	$-\frac{1}{3}$	$-\frac{1}{15}$	$-\frac{1}{35}$	$-\frac{1}{63}$	$-\frac{1}{99}$
1	0	$-\frac{1}{3}$	$\frac{4}{15}$	$-\frac{1}{35}$	$\frac{16}{315}$	$-\frac{1}{99}$
2	0	$\frac{1}{15}$	$\frac{4}{21}$	$-\frac{5}{21}$	$\frac{16.17}{5.7.9.11}$	$-\frac{673}{7.9.11.13}$
3	0	0	$-\frac{4}{35}$	$-\frac{8}{3.5.7}$	$\frac{16.41}{5.7.9.11}$	$-\frac{8.107}{7.9.11.13}$
4	0	0	$\frac{4}{315}$	$\frac{8.19}{3.5.7.11}$	$-\frac{16}{7.11.13}$	$-\frac{8.47}{3.7.11.13}$
5	0	0	0	$-\frac{8}{231}$	$-\frac{17.64}{7.9.11.13}$	$\frac{8}{99}$

$$\lambda = 2$$

n \ m	0	1	2	3	4	5
0	$\frac{1}{2}$	0	$-\frac{1}{3}$	0	$-\frac{1}{15}$	0
1	0	$-\frac{1}{5}$	$\frac{4}{105}$	$\frac{11}{105}$	$-\frac{16}{11.105}$	$\frac{97}{11.13.21}$
2	0	$\frac{1}{45}$	$\frac{32}{315}$	$-\frac{17}{315}$	$-\frac{8.16}{5.7.9.11}$	$\frac{61}{7.9.11.13}$
3	0	0	$-\frac{2}{63}$	$\frac{4}{77}$	$\frac{8.17^2}{5.7.9.11.13}$	$\frac{4.17}{7.9.11.13}$
4	0	0	$\frac{4}{5.5.7.9}$	$\frac{64}{5.5.7.11}$	$\frac{16.59}{5.7.9.11.13}$	$-\frac{64.29}{5.7.9.11.13}$
5	0	0	0	$-\frac{8}{9.11.13}$	$-\frac{64}{3.5.11.13}$	$-\frac{8}{3.11.13.17}$

$$\lambda = 3$$

n \ m	0	1	2	3	4	5
0	$\frac{1}{2}$	$\frac{1}{5}$	$-\frac{13}{35}$	$-\frac{11}{105}$	$-\frac{61}{11.105}$	$-\frac{97}{11.13.21}$
1	0	$-\frac{2}{15}$	$-\frac{4}{105}$	$\frac{2}{21}$	$\frac{16}{11.105}$	$\frac{46}{11.13.21}$
2	0	$\frac{1}{105}$	$\frac{2}{35}$	$-\frac{1}{11.105}$	$-\frac{8.71}{11.13.105}$	$-\frac{4}{11.13.21}$
3	0	0	$-\frac{2}{5.35}$	$-\frac{4.41}{55.105}$	$\frac{8.19}{105.143}$	$\frac{4.61}{105.143}$
4	0	0	$\frac{4}{55.105}$	$\frac{8.101}{5.105.143}$	$\frac{16}{11.105}$	$-\frac{8.349}{17.105.143}$
5	0	0	0	$-\frac{32}{7.21.143}$	$-\frac{64}{49.143}$	$-\frac{8.12}{7.17.143}$

Tables of the weight coefficients $\rho_j^{(\lambda,m)}$ in Equation (2.25), as defined by Equation (2.27), for $\lambda = 1, 3, 5$, and $j = 1(1)4$ and $n = 0(1)5$.

$$\lambda = 1$$

n	j 1	2	3	4
0	$\frac{1}{2}$	$\frac{1}{8}$	$\frac{1}{16}$	$\frac{5}{128}$
1	$-\frac{1}{2}$	$\frac{1}{8}$	$\frac{1}{16}$	$\frac{5}{128}$
2	$\frac{1}{2}$	$-\frac{7}{8}$	$\frac{3}{16}$	$-\frac{11}{128}$
3	$-\frac{1}{2}$	$\frac{11}{8}$	$-\frac{9}{16}$	$-\frac{25}{128}$
4	$\frac{1}{2}$	$-\frac{19}{8}$	$\frac{41}{16}$	$-\frac{15}{128}$
5	$-\frac{1}{2}$	$\frac{29}{8}$	$-\frac{111}{16}$	$\frac{425}{128}$

$$\lambda = 3$$

n	j 1	2	3	4
0	$\frac{3}{2}$	$-\frac{3}{8}$	$-\frac{1}{16}$	$-\frac{3}{128}$
1	$-\frac{1}{2}$	$\frac{5}{8}$	$-\frac{1}{16}$	$-\frac{3}{128}$
2	$\frac{1}{4}$	$-\frac{11}{16}$	$\frac{13}{32}$	$\frac{9}{256}$
3	$-\frac{3}{4.5}$	$\frac{3.19}{4.20}$	$-\frac{11.13}{4.40}$	$\frac{3.19}{4.64}$
4	$\frac{1}{2.5}$	$-\frac{29}{2.20}$	$\frac{11.11}{2.40}$	$-\frac{3.43}{2.64}$
5	$-\frac{1}{14}$	$\frac{41}{14.4}$	$-\frac{11.13}{14.8}$	$\frac{3.787}{14.64}$

$$\lambda = 5$$

n	j 1	2	3	4
0	$\frac{5}{2}$	$-\frac{15}{8}$	$\frac{5}{16}$	$\frac{5}{128}$
1	$-\frac{1}{2}$	$\frac{9}{8}$	$-\frac{11}{16}$	$\frac{5}{128}$
2	$\frac{1}{6}$	$-\frac{7}{6.4}$	$\frac{43}{6.8}$	$\frac{5.25}{6.64}$
3	$-\frac{1}{14}$	$\frac{27}{14.4}$	$-\frac{113}{14.8}$	$\frac{5.139}{14.64}$
4	$\frac{1}{28}$	$-\frac{39}{28.4}$	$\frac{241}{28.8}$	$-\frac{5.487}{28.64}$
5	$-\frac{5}{63.4}$	$\frac{5.53}{63.16}$	$-\frac{5.11.41}{63.32}$	$\frac{5.61.109}{63.256}$

VI.5 Bibliographical Notes

The idea of "telescoping" of Fourier cosine series which represent the light variations of close binary systems caused by the distortion of figure, goes back to Sec. VI.12 of Kopal (1959); and was developed further in Kopal (1975, 1976) and subsequent literature.

The process of "modulation" of the light curves to remove the effects of distortion was initiated by Kopal (1976), and developed further in 1982c and, in more practical terms, in Sec. 6.3 of Kopal (1986).

A systematic investigation of the "photometric perturbations" $\mathcal{P}_{2m}$ on the respective moments of the light curves goes back to Kopal (1975d), and was subsequently developed in more detail by Livaniou (1977), Rovithis Livaniou (1978, 1979) and Niarchos (1976) in the course of their respective doctoral research. For later work the reader may be referred to Kopal (1982c and 1987); though the material presented in Sec. VI.3 of the present book is mostly new.

Chapter VII

LABORATORY SIMULATIONS

In the principal part of this book a new mathematical theory of the optical phenomena underlying mutual eclipses of the components of close binary systems—initiated by the present writer in 1977—has been outlined in its present state. This theory was originally developed for the sake of astrophysical applications to double-star astronomy, and reached a state far exceeding the previous stage of the subject. However, an account of these would not be complete if an explicit mention was not made of the fact that applications of this theory are not necessarily limited to the field of astronomy, but can be developed further to serve also other branches of human science or technology; and the principal one of these appears to be in the field of *automatic computers*.

The impact of such devices, in recent decades, on a vast domain of human endeavour—including astronomy—needs indeed no particular emphasis in this place; but a few words concerning the history of this subject may not be amiss. This is especially true of any book devoted to astronomy or mathematics from time immemorial—remember the inscription on the entrance of the Academy of divine Platon (of which our more modern universities can be regarded as distant descendants), "Let No One Enter this Gate Who is not Conversant with Mathematics". Mathematics meant, in those days, largely a study of the motions of the celestial bodies in the sky. Not perhaps, for Platon himself—who was interested more in appearances than realities—but certainly to Aristotle, for whom the regular motions of celestial bodies in the sky were a sufficient proof of the existence of God.

It is indeed interesting to consider which course the history of mathematics might have taken if our Solar System possessed only one planet—our Earth—or if the Earth possessed no Moon. For it was these "wandering stars" which confronted the mathematicians with ever-increasing challenges in the sky—from the Pythagoreans through Eudoxos to Hipparchos and Ptolemy. The celebrated theory of homocentric spheres and epicyclic motions of antiquity represent, in effect, nothing else but a geometric representation of Fourier expansions of the planetary theory deduced from the Newtonian mechanics!

These geometrical theories had, to be sure, still little to do with numerical analysis. The latter's cradle stood in Mesopotamia, and its inspiration was only in part astronomical. Although the base of 10—until quite recently the most widespread basis of numerical computations—did not come to us from Babylonia (it was used already in Egypt; the Babylonians were wedded to the base of 60 mainly for calendrical reasons); it was the Babylonians who introduced the concept of the *positional meaning* of numbers; while for the third great discovery which made our numerical analysis what it is today—namely, that of *zero*—we must go to India many centuries later. It was indeed the Hindus who invented the system known to us as the "Arabic numerals"—called so before the medieval Arabs transmitted its discovery to the West.

However, since the time of the Renaissance numerical analysis was developed primarily by astronomers, and for very good reasons. For as early as in the 17th century astronomy became the first branch of science to come into possession of the natural law of universal gravitation—of far-reaching exactitude for its simple form—which permitted accurate deductions to be made from it by analysts of the calibre of Newton, Euler, Lagrange, Gauss or Poincaré (to name only the greatest) whose work profoundly influenced the whole development of mathematics. Moreover, by way of a parallel effort, the techniques of positional measurements had been developed by astronomers to carry out comparisons between theory and observations with accuracy unattainable in any other branch of science. To do so required, in turn, a development of numerical processes adequate to carry out such comparisons in a systematic manner. This is why theoretical astronomy became the cradle of numerical analysis as we know it today; and the names of the principal contributors to the mathematical armoury required to this end—in the form of interpolation or quadrature formulae, least-squares techniques, etc.— like Newton himself, Laplace, Gauss, Bessel and others—read like a part of the history of astronomy as well as mathematics of their age.

In more recent times, however, the event which influenced contemporary numerical analysis even more than celestial bodies has been a development of *computing machines* which have given an entirely new dimension to our effort for quantitative studies of natural phenomena, unthinkable only a generation or so ago. The pedigree of such devices—going back to Leibniz or Pascal in the 17th century, or Babbage and Kelvin in the 19th—is almost as old as that of the infinitesimal calculus itself; but their means of attaining our ends may be very different.

In order to explain the circumstances, let us stress that the introduction of infinitesimal calculus in applied mathematics at the end of the 17th century, usually regarded as the primary cause of this trend of events, did represent, to be sure, an intellectual triumph unparalleled in the history of human endeavour. However—in retrospect—these triumphs should not be allowed to conceal the fact that they were limited largely to those branches of science in which natural phenomena are controlled by differential equations of *integral* orders, that are *linear* in their dependent variables. This is true (as is well known) of Maxwell's equations of electrodynamics, or of Schrödinger's equation at the basis of elementary quantum mechanics; and in these fields a wide class of problems expressible in terms of such equations have been solved analytically in the grand manner.

On the other hand, in the domains of science describable in terms of nonlinear equations (of integral orders)—such as mechanics (both classical and relativistic) or hydrodynamics—the methods of formal analysis have failed so far to provide the general answers that were expected of them. Consider, for instance, the equations governing the motions of the celestial bodies in space, which consist of a simultaneous system of ordinary differential equations of second order. These equations are, however, nonlinear in spatial coordinates which represent their dependent variables; and as such have not only defied integration by known methods in a finite number of terms; but the validity of infinite expansions aiming to represent their integrals are still under discussion. In the case of hydrodynamics, the situation becomes still very much more complicated because of *discontinuities* inherent in their solutions; so much so that a knowledge of the form of nonlinear partial differential equations, used for their description, scarcely lessens our ignorance of the phenomena which such equations attempt to describe.

In order to penetrate this ignorance, numerical analysts of the past half a century have considered a return to Fermat (i.e., pre-Newtonian) days by a systematic *algebraization of analysis* by replacing the derivatives (fluxions) by finite differences—an approach in which a futile quest of general analytic solutions has been replaced by the construction of particular solutions of the respective problems by numerical methods. In such a "calculus of finite differences" many of the headaches of infinitesimal analysis—such as the convergence of infinite series—can indeed be neatly avoided. For whenever we deal with a sequence of numbers rather than of algebraic symbols, we always deal with polynomials which, for finite values of the independent variable are always bounded; what matters then alone are asymptotic properties of the respective polynomials rather than convergence of the respective infinite series.

On the other hand, other dangers of such an approach lurk in the background. For the process of an algebraization of analysis, differential equations (or systems) of n-th order is being replaced by an equivalent system of difference equations of order generally much greater than n for simulation to be sufficiently effective. Such systems admit, however, also of many more particular solutions than the original system of equations; and their admixture (never completely suppressible by imposition of boundary conditions which can be numerically spelled out only to a finite number of digits) may eventually overwhelm the outcome—not to speak of the effects caused by an accumulation of round-off errors inherent in any digital work.

VII.1   Analog Computers: Will Digital Computation Survive?

In the face of such a situation, an alternative approach to digital computation has been initiated since the 1930's with the development of "analog-type" computing devices which do not actually compute, but measure the dependent variables under conditions simulating the solution of a given problem. In this sense—for instance—a wind tunnel could be regarded as an analog-type "computer" solving the differential equations of certain types of fluid flow; though we do not call it a computer because it can perform this task for only a very restricted class of problems (cf., e.g., Thomson, 1876).

More general types of analog computers were, however, developed in the 1930's by Vannevar Bush and his associated at M.I.T. in Cambridge, Mass., and continued by D. R. Hartree at the University of Manchester in England (for their description cf. Hartree, 1949), based on—largely on account of the "state of the art" of contemporary technology—by measurement of the angles and lengths (or, at a somewhat later stage, of voltage). The elementary operations which these devices were called upon to perform were "integration" (i.e., planimetry) and "averaging", from which all other and digitally more elementary algebraic operations (such as addition, subtraction, etc.) had to be extracted—multiplication or division of two functions u and v, for instance, from the well-known formulae

$$uv = \int u'v + \int uv' \tag{1.1}$$

or

$$\frac{u}{v} = \int \frac{u'}{v} - \int \frac{u}{v}\frac{v'}{v} \tag{1.2}$$

by partial integration.

The accuracy attainable by such processes depended essentially on the precision with which the measuremens of the quantities underlying their logic could be carried out—approximately 1 part in 10^3 for the early Bush-type mechanical analyzer of the 1930's, increased about 10 fold in 1943 with the advent of the Mark II Rockefeller differential analyzer with numerical output. But—and this was essential—the individual elements of such systems could work only in parallel rather than serially, and thus were unable to *compound accuracy*—in contrast with digital computing machines, for which the accuracy of a (say) 10-digit computer can be increased ten-fold by adding one extra column of decimals to their register. It is the ability to compound accuracy in this way that gained for the digital computers the ascendancy in the past half a century, since the first (relay-type) computers—slow and clumsy—were replaced by those working internally at electronic speeds, the efficiency of which became limited only by their input-output requirements.

Their ascendancy could, however, come very quickly to an end, once technical ways have been discovered to make the individual elements of analog computers work serially rather than in parallel. If, for example, an analog-type integrator—accurate on its own to a precision of (say) 1 part in 10^4—could be made to operate within a 10^{-4}th part of the error of another integrator coupled serially with it, an accuracy of 1 part in 10^8 could immediately be attained; and if so, there is no doubt that such combinations could displace digital computers out of practical use in a very short time.

There are several obvious advantages which should enable them to do so—especially since analog devices need not be impeded in their performance by eventual non-linearities of the underlying mathematical problems, nor of their possible singularities; for physical measurements do not suffer from such limitations; and nature abhors them.

Like all physical devices, analog computers are bound to be subject to specific types of errors arising from the uncertainty of measurement of the quantities which its elements may be called upon to perform. The most precise measurements of which physical science is capable at the present time are not those of the lengths or angles (or of voltages) on which analog computers were based in the past, but of the *time* (or, of its reciprocal, *frequency*). While the lengths can still be measured in the laboratory with an engineering accuracy of not much less than 1 part in 10^4, the time (or frequency) is being currently measured to a precision of the order of 1 part in 10^{13} by atomic clocks over time-intervals which are long when compared with the duration of the respective computation; and a serial combination of n devices of this type could (theoretically) attain an accuracy of 1 part in 10^{13n}.

The question immediately arises: what kind of measuring devices could be thought of to attain an accuracy of this order? The answer is clear: such devices would have to be *optical*—with photons taking over the role of the electrons in contemporary digital computers—and compounding of the accuracy should be accomplished by interference of coherent light beams; the limitation of the round-off errors in digital computers being superseded by those inherent in the quantum nature of light.

The use of optical analogs in computers of the future will require also a reformulation of mathematical problems to be solved conveniently with their aid. Because of the ease with which electronic circuits can be made to count, digital computers require the problems to be served to them in arithmetic form; and the difficulties inherent in an "algebraization of analysis" have already been touched upon earlier in this chapter. On the other hand, a simulation of analytical problems by optical analogs leads—by the very nature of physical optics—to a treatment of such problems in the *frequency-domain*: the unknown functions being generated internally through the intermediary of their Fourier (or other integral) transforms.

The idea to use such transforms in analog computers is not new: its elements were already built in the mechanical integrators of the "differential analyzers" of the 1930's. But it will not be until the advent of optical analog computers of more general type (for which the "spectral analysis" constitutes their natural language) that this technique really comes into its own. As was already mentioned, digital computation represents a basically clumsy process for coping with infinitesimal analysis (of which the "calculus of finite differences" is only a half-legitimate stepchild).

It is, however, well known that, in the frequency-domain, finite differences possess *continuous* Fourier transforms; and, as such, can be dealt with by continuous processes not requiring any "deferred approach to the limit". In point of fact, virtually *the entire scope of the calculus of finite differences can be represented by continuous processes in the frequency-domain*; and the Fourier transforms of wide classes of functions are actually simpler than the parent functions which can be synthesized from them. This constitutes the main attraction which optical analog computers offer in the future; and their emergence as a leading tool for quantitative exploration of the physical world may be around the corner.

And more: for in the present book, devoted primarily to mathematical theory of stellar eclipses, we already came across some instances of such a situation. Consider, for example, Equation (1.34) of Chapter III, expressing the fractional loss of light α_ν during eclipses of spherical stars as a Hankel transform

$$\alpha_n^0 = \int_0^\infty \mathcal{A}_\nu(x)\, J_0(hx)\, x\, dx \,, \tag{1.3}$$

of zero order (where $\nu \equiv (n+2)/2$), of the product

$$\mathcal{A}_\nu(x) \equiv 2^\nu \Gamma(\nu) \frac{J_\nu(kx)}{(kx)^\nu} \frac{J_1(x)}{x} \tag{1.4}$$

of two Bessel functions of arbitrary parameters $h \equiv \delta/r_2$ and $k \equiv r_1/r_2$, as defined by Eqs. (1.35) of Chapter III.

The inverse of the Hankel transform (1.3) can be expressed as

$$\mathcal{A}_\nu(x) = \int_0^\infty \alpha_n^0 J_0(hx)\, h\, dh \equiv$$

$$\equiv \int_{h_0}^{h_1} \alpha_n^0 J_0(hx)\, h\, dh + \int_0^{h_0} \alpha_n^0 J_0(hx)\, h\, dh \,, \tag{1.5}$$

in which the limits of the first integral can be restricted from (∞, h_0) to (h_1, h_0), where h_0 stands for the phase of maximum eclipse and h_1, for that of its first contact. To do so is indeed legitimate since, for $h > h_1$, $\alpha_n(h, k) = 0$; and, moreover, the value of the integral between (h_1, h_0) can be ascertained by a planimetry of the observed light changes as explained in Chapter V in terms of the "moments of the light curves" A_{2m} as defined by Eq. (1.1) of Chapter V. For the second integral on the r.h.s. of Eq. (1.5) this is, however, not true any more; as the light changes for $h < h_0$ are no longer observable; and to render them such would require tilting of the orbital plane of our binary to a position perpendicular to the celestial sphere.

What cannot be done in the sky can, however, be accomplished readily enough in the laboratory, where any value of h_0 including zero is entirely legitimate for the purpose of simulation. By setting $h_0 = 0$, however, *the inverse Hankel transform $\mathcal{A}_\nu(x)$ of α_n^0—i.e., a product of two Bessel functions of arbitrary arguments—can be evaluated in terms of the moments A_{2m} appropriate for a central eclipse*, the values of which can be ascertained by existing means within an error of about 1 part in 10^4—i.e., much more accurately than from astronomical observations in the sky (for the laboratory light beam can be arbitrarily bright; with intensity unimpaired by the vagaries of atmospheric extinction).

In this way, the products of Bessel functions of arbitrary order—not easy to evaluate numerically for large values of the respective arguments (at least by the

1944 Harvard Mark I relay-type calculator which spent most of its useful lifetime on such tasks)—can be established also by photometric means. Products of Bessel functions—encountered as they often are in many tasks arising in electronics or electrical engineering, as well as in other aspects of human endeavour—are certainly of considerable practical interest.

Laboratory photometry outlined in this section to this end does not yet entitle the respective device to be regarded as a "computer"—any more than it would be to consider the wind-tunnel a potential solver of partial differential equations. However, the role of the former is readily susceptible of considerable generalization. While in all astronomical situations considered in Chapters IV-VI both components of the eclipsing binary systems have been assumed to possess shapes appropriate for hydrostatic equilibrium in the prevalent field of force, the optical masks simulating such eclipses can be made, first, of arbitrary shape to suit the requirements of the respective mathematical problem rather than any physical situations; and, secondly, the time-scale of laboratory simulations is not set by the celestial phenomena, but by the experimental means employed.

In this way, optically-monitored laboratory simulation of eclipse phenomena can be brought to bear on a very wide class of mathematical problems—linear or nonlinear—not easily amenable to digital computation. A more detailed exposition of such an approach would, however, transcend by far the scope of the present monograph devoted essentially to more specific astronomical implications, and must be deferred to a separate volume to appear in the future. But whatever the case may be, there is no room for doubt that a set of accurately measured physical quantities can, in principle, be exchanged for another model with the same variables, but of different information content, which can determine their value as a "computer"; and among these the measurements (cf. Gray, 1931; or Hazen and Brown, 1940), which Hartree referred to as "photoelectric integraphs"—represented in the sky by our eclipsing variables—can, in the future, hold an especially prominent place. In fact, our present work goes beyond that of Gray (referred to above) in two respects: (a) our solution of the problem as given in Chapter III involves two parameters (i.e., h and k) rather than the one employed by Gray; and (b) it makes explicit use of Hankel, rather than Fourier, transforms. But otherwise our general approach to the problem continues to remain the same; and may be extended in the future to the entire work summarized in this book.

REFERENCES

Appell, P.: 1880, Ann. de l'Ecole Normale Paris, (2)**9**, 119.

Appell, P. and Kampé de Fériet, J.: 1926, *Fonctions Hypergéometriques et Hyperspheriques*, Gauthier-Villars, Paris.

August, A.: 1881, Archiv Math. Phys., **67**, 72.

Bailey, W.N.: 1935, *Generalized Hypergeometric Series*, Cambridge Tracts in Mathematics and Mathematical Physics No. 32, Cambridge Univ. Press.

Bailey, W. N.: 1936, Proc. London Math. Soc., **40**, 37.

Bateman, H.: 1905, Proc. London Math. Soc., **3**, 111.

Binnendijk, L.: 1966, Astron. J., **71**, 340.

Bruns, H.: 1882, Monatsber. d. Preuss. Akad. d. Wiss. für 1881, p.48ff.

Budding, E.: 1988, Southern Stars, **32**, (No.6), 180.

Burchnall, J. L.: 1942, Quart. J. Math. (Oxford) **13**, 90.

Bush, V.: 1931, J. Franklin Inst., **212**, 447.

Chaundy, T. W.: 1943, Quart. J. Math. (Oxford), **14**, 55.

Christoffel, E.B.: 1858, J. Reine Angew. Math., **55**, 61.

Clegg, F. E.: 1962, The School Science Rev., No.147.

Cox, J. P.: 1980, *Theory of Stellar Pulsation*, Princeton Univ. Press, Princeton, N.J.

Demircan, O.: 1978, Astrophys. Space Sci., **59**, 313.

Erdélyi, A., Magnus, W., Oberhettinger, F. and Tricomi, F. G.: 1954, *Tables of Integral Transforms*, vol. II, McGraw Hill Book Co., New York.

Fracastoro, M. G.: 1972, *Atlas of the Light Curves of Eclipsing Binaries*, Torin.

Goodricke, J.: 1783, Phil. Trans. Roy. Soc. London, **73**, 474.

Gray, T. S.: 1931, J. Franklin Inst., **212**, 77.

Hartree, D. R.: 1949, *Calculating Instruments and Machines*, Univ. of Illinois Press, Urbana.

Hazen, H. L. and Brown, G. S.: 1940, J. Franklin Inst., **230**, 19, 183.

Henn, K.: 1889, Math. Ann., **33**, 161.

Hoskin, M.: 1979, J. History of Astron., **10**, 23.

Humbert, P.: 1924, J. de l'Ecole Polytech., (2)**24**, 59.

Kitamura, M.: 1967, *Tables of the Characteristic Functions of the Eclipse and the Related Delta Functions for the Solution of Light Curves of Eclipsing Binary Systems*, Univ. of Tokyo Press.

Kopal, Z.: 1941, Astrophys. J., **94**, 159.

Kopal, Z.: 1942a, Astrophys. J., **96**, 20.

Kopal, Z.: 1942b, Proc. Amer. Phil. Soc., **85**, 399 (Harvard Reprint Ser. II, No. 1).

Kopal, Z.: 1942c, Proc. U.S. Nat. Acad. Sci., **28**, 133.

Kopal, Z.: 1945, Proc. Amer. Phil. Soc., **89**, 517.

Kopal, Z.: 1947, Harvard Obs. Circ., No.450.

Kopal, Z.: 1949, Harvard Obs. Circ., No.454.

Kopal, Z.: 1959, *Close Binary Systems*, Chapman-Hall and John Wiley, London and New York.

Kopal, Z.: 1961, *Numerical Analysis* (2nd ed.), Chapman-Hall and John Wiley, London and New York.

Kopal, Z.: 1975a, Astrophys. Space Sci., **34**, 431.

Kopal, Z.: 1975b, Astrophys. Space Sci., **35**, 159.

Kopal, Z.: 1975c, Astrophys. Space Sci., **35**, 171.

Kopal, Z.: 1975d, Astrophys. Space Sci., **38**, 191.

Kopal, Z.: 1976, Astrophys. Space Sci., **45**, 269.

Kopal, Z.: 1977a, Astrophys. Space Sci., **46**, 87.

Kopal, Z.: 1977b, Astrophys. Space Sci., **50**, 225.

Kopal, Z.: 1977c, Astrophys. Space Sci., **51**, 439.

Kopal, Z.: 1978, *Dynamics of Close Binary Systems*, D. Reidel Publ. Co., Dordrecht and Boston.

Kopal, Z.: 1979, *Language of the Stars*, D. Reidel Publ. Co., Dordrecht and Boston.

Kopal, Z.: 1982a, Astrophys. Space Sci., **81**, 123.

Kopal, Z.: 1982b, "Contribution to Fourier Analysis of the Light Curves of Eclipsing Variables" in (E. G. Mariopoulos, Ed.) *Compendium in Astronomy* (D. Reidel Publ. Co., Dordrecht and Boston); p.233ff.

Kopal, Z.: 1982c, Astrophys. Space Sci., **81**, 411.

Kopal, Z.: 1982d, Astrophys. Space Sci., **87**, 149.

Kopal, Z.: 1982e, Astrophys. Space Sci., **88**, 313.

Kopal, Z.: 1983, Astrophys. Space Sci., **90**, 445.

Kopal, Z.: 1986a, Vistas in Astronomy, **29**, 295.

Kopal, Z.: 1986b, *Of Stars and Men (Reminiscences of an Astronomer)*, Adam Hilger Ltd., Bristol and Philadelphia; Chapter VI.

Kopal, Z.: 1987, Astrophys. Space Sci., **134**, 235.

Kopal, Z.: 1989, *The Roche Problem and its Significance in Double-Star Astronomy*, Kluwer Academic Publishers, Dordrecht and Boston.

Kopal, Z., Plavec, M. and Reilly, E. F.: 1960, Jodrell Bank Annals, **1**, 374–423.

Kopal, Z. and Demircan, O.: 1978, Astrophys. Space Sci., **55**, 241.

Kopal, Z. and Yamasaki, A.: 1980, Astrophys. Space Sci., **72**, 3.

Kukarkin, B. V. and Parenago, P. P.: 1948, *Obschii Katalog Peremennyich Zvjozd*, Publ. Sternberg State Astr. Inst., Moscow.

Lanzano, P.: 1976a, Astrophys. Space Sci., **42**, 425.

Lanzano, P.: 1976b, Astrophys. Space Sci., **45**, 419.

Lanzano, P.: 1976c, Astrophys. Space Sci., **45**, 483.

Livaniou, H.: 1977, Astrophys. Space Sci., **51**, 77.

Livaniou-Rovithis, H.: 1977, Astrophys. Space Sci., **52**, 271.

Livaniou-Rovithis, H.: 1978, Astrophys. Space Sci., **59**, 463.

McLaughlin, D.B.: 1924, Astrophys. J., **60**, 22.

Milne-Thomson, L. M.: 1933, Proc. London Math. Soc., (2)**35**, 514.

Montanari, G.: 1671, "Sopra la Sparizione d'Alcune Stelle e Altre Novità Celesti", in Prose di Signori Academici Gelati di Bologna.

Niarchos, P.: 1978, Astrophys. Space Sci., **58**, 301,

Oberhettinger, F.: 1973, *Fourier Expansions*, Academic Press, New York and London.

Olson, E. C.: 1976a, Astrophys. J., **204**, 141,

Olson, E. C.: 1976b, Astrophys. J. Suppl., **31**, 1–11.

Olson, E. C.: 1982, Publ. Astr. Soc. Pacific, **94**, 70.

Petrie, R. M.: 1938, Publ. Dominion Astrophys. Obs., **7**, 133.

Pickering, E.C.: 1880, Proc. Amer. Acad. Arts. Sci., **16**, 1.

Porro, A.: 1891, Astron. Nachr., **127**, 41.

Radau, R.: 1880, J. de Math. (3)**6**, 283.

Rice, S. O.: 1935, Quart. J. Math. (Oxford), **6**, 52.

Rossiter, R. A.: 1924, Astrophys. J., **60**, 15.

Russell, H. N.: 1906, Astrophys. J., **24**, 1.

Schlesinger, F.: 1909, Allegheny Obs. Publ., **1**, 126.

Schwab, F.: 1901, Astron. Nachr., **157**, 79.

Slater, L. J.: 1966, *Generalized Hypergeometric Functions*, Cambridge Univ. Press.

Srivastava, H. M. and Kopal, Z.: 1989, Astrophys. Space Sci., **162**, 181.

Tassoul, J. L.: 1978, *Theory of Rotating Stars*, Princeton Univ. Press.

Thomson, W. (Lord Kelvin): 1876, Proc. Roy. Soc. London, **24**, 269 and 271.

Tricomi, F. G.: 1935, Rendi Conti dei Lincei, (6)**22**, 564.

Truesdell, C.: 1948, *A Unified Theory of Special Functions* (Annals of Math. Studies, No. 18), Princeton Univ. Press.

Tsesevich, V. P.: 1936, Publ. Univ. Obs. Leningrad, **6**, 48 (in Russian).

Tsesevich, V. P.: 1939, Bull. Astr. Inst USSR Acad. Sci.,No. 45.

Unno, W., Osaki, Y., Ando, H. and Shibasaki, H.: 1979, *Nonradial Oscillations of Stars*, Univ. of Tokyo Press.

Vogel, H. C.: 1890, Astron. Nachr., **123**, 289.

Watson, G. N.: 1945, *A Treatise on the Theory of Bessel Functions* (2nd ed.), Cambridge Univ. Press.

Whittaker, E. T. and Watson, G. N.: 1920, *Modern Analysis* (3rd ed.), Cambridge Univ. Press.

Zhevakin, S. A.: 1953, Astron. Zhurnal, **30**, 161.

Zhevakin, S. A.: 1954, Astron. Zhurnal, **31**, 141, 335.

INDEX